# The Essential
# Howard Gardner
# on Mind

Howard Gardner

TEACHERS COLLEGE PRESS

TEACHERS COLLEGE | COLUMBIA UNIVERSITY
NEW YORK AND LONDON

Published by Teachers College Press,® 1234 Amsterdam Avenue, New York, NY 10027

Front cover art by Jay Gardner.

*Library of Congress Cataloging-in-Publication Data*

Names: Gardner, Howard, 1943- author.
Title: The essential Howard Gardner on mind / Howard Gardner.
Description: New York, NY : Teachers College Press, [2024] | Includes bibliographical
    references and index. | Summary: "Howard Gardner's life's study and his theory of
    multiple intelligences show how an understanding of human cognitive capacities
    and processes manifests itself in several domains, such as artistry, leadership,
    creativity, and excellence in the professions"— Provided by publisher.
Identifiers: LCCN 2024013606 (print) | LCCN 2024013607 (ebook) |
    ISBN 9780807769362 (paperback ; alk. paper) | ISBN 9780807769379
    (hardcover ; alk. paper) | ISBN 9780807782224 (ebook)
Subjects: LCSH: Cognition. | Developmental psychology. | Multiple intelligences. |
    Psychology—Philosophy.
Classification: LCC BF311 .G338 2024 (print) | LCC BF311 (ebook) |
    DDC 153—dc23/eng/20240510
LC record available at https://lccn.loc.gov/2024013606
LC ebook record available at https://lccn.loc.gov/2024013607

ISBN 978-0-8077-6936-2 (paper)
ISBN 978-0-8077-6937-9 (hardcover)
ISBN 978-0-8077-8222-4 (ebook)

Printed on acid-free paper
Manufactured in the United States of America

# Contents

## POSITIVE USES OF MIND: INTRODUCTION TO GOOD WORK

## THE PROFESSIONS

# Acknowledgments

In over a half-century of research and writing, I've received wonderful help from hundreds of individuals—teachers, peers, students, editors, members of my research team, colleagues from institutions spread across the globe—and, importantly, many generous funding agencies. When possible, I have acknowledged their help directly in these writings. But all too often, this has not proved possible. And so, with awareness of the inadequateness of the response, I voice my sincere gratitude to each and every one of you—invisible collaborators, so to speak.

In arranging for the publication of these volumes, I am tremendously indebted to Teachers College Press (TCP) at Columbia University. Brian Ellerbeck recognized the potential of this ambitious publishing venture and has been a true thought partner in the evolution from drafts of too many candidate pieces to the selection and presentation of the essays in these two volumes. Thank you, Brian!

In addition, I am grateful to production editor Mike Olivo; copyeditor Pam LaBarbiera; proofreader Liz Welch; and their many responsible and responsive colleagues at TCP.

Only someone who has attempted to identify, track down, and locate the appropriate personnel, and secure permissions from many outlets scattered around the world, can appreciate the heroic work done by two unbelievably gifted and devoted colleagues, Shinri Furuzawa and Annie Stachura. In addition, Annie and Shinri have worked with me on countless details—not only taking care of them deftly, but calming me down when that was indicated.

Last, but also first, a profound thank-you to my family. Above all to my wife, Ellen Winner, who has participated in every facet of this undertaking—problem-finding, problem-solving, and holding my hands, both literally and figuratively. And while they were largely spared any requests associated with these two volumes, my children, Kerith, Jay, Andrew, and Benjamin, have supported me over the decades—whether it was managing to fall asleep while I was typing late into the night, or giving me an idea for a study, or providing a critique—or simply a hug. I hope that if anyone asks them what their father accomplished in his research and writings, they will point to *The Essential Howard Gardner*.

# Introduction

When I was a teenager, my Uncle Fred gave me a textbook in psychology. I had probably heard the word "psychology" before, but I had essentially no idea about the field of study that it denoted. But Fred was shrewd. He sensed that I was interested in the operations of the human mind, though at the time it was not an interest that I was consciously aware of.

To be sure, I knew that I was interested in people and how they lived, worked, played, struggled, and thought. I came from a household that was rich in conversation about human beings—their experiences, aspirations, motivations, challenges, fates. And as an engaged learner, I read widely about human beings, especially history and biography. And when I read novels or stories, I was much less interested in the settings or even the uses of language than in the biographical information, personalities, and thought processes of the protagonists.

So it was hardly a surprise that when I got to college, after a flirtation with the study of history, I elected to "concentrate" in a field of study called "Social Relations," which spanned psychology, sociology, and anthropology. I enjoyed this interdisciplinary work; in fact, my undergraduate thesis—a field study of what I dubbed Leisure Village, one of the first retirement communities in the United States—drew on all three of these human-oriented disciplines.

I was uncertain about what to do after college. That state of mind evaporated after a summer job working with Jerome Bruner, one of the founders of the emerging field of cognitive psychology. This foray into one branch of "social relations" convinced me to pursue doctoral studies in psychology, and to focus on what came to be known as cognitive development—a field of research that had been essentially launched early in the 20th century by the Swiss psychologist Jean Piaget and of which Bruner was arguably the principal proponent in the United States.

Because I was always a bit of an intellectual sprawler (some might quip "dilettante"), in graduate school I was also influenced by a wide number of scholars—ranging from anthropologist Claude Lévi-Strauss to philosopher Nelson Goodman, from art historian Ernst Gombrich to neurologist Norman Geschwind. Unlike most of my fellow graduate students in psychology, I discovered that I was more of a book writer than an author of articles for scholarly journals; and that I tended to stretch across fields and methods rather than to focus on a particular problem using a sanctified research method—the preferred

1

route for someone aiming to get a job as an assistant professor in a department of psychology (or, indeed, in *any* academic department).

Enough autobiography. For over half a century, I have been interested in the human mind—in its various shapes, forms, and operations. Empirically, I have studied how children's minds develop, particularly in the arts (developmental psychology); and, as a complement, I have examined how cognition breaks down as a consequence of damage to the brain (neuropsychology). As a result of my daily research with these two populations, I arrived around 1980 at the body of work for which I am best known—the theory of multiple intelligences. The attention paid to that work resulted in much discussion and debate with colleagues, as well as a lengthy foray into education—one fully documented in the accompanying volume (*The Essential Howard Gardner on Education*).

Dating back to my youth, I have been fascinated by the heights of intellect, artistry, creativity, or political leadership achieved by a few members of our species. (The portraits hung in my childhood bedroom were of Albert Einstein and Ernest Hemingway, and I cherish a 1956 newspaper photo where I was shooting a movie clip of then–presidential candidate Adlai Stevenson). Along the way, I have also studied certain special human cognitive capacities, chief among them the capacity to synthesize, and influences on cognition, especially those afforded by new media and technology. And, living in an age of constant change, even turmoil, I've thought about the kinds of minds that are likely to be most valuable in the future.

In recent decades, my colleagues and I have focused on roles of crucial importance in our world—particularly the roles of worker and of citizen; and we have sought to understand and to promote the positive potentials of our species—humankind—what we've termed good work, good play, good collaboration, and good citizenship. This panoramic stance takes me back to my original work in "social relations": I draw on the tools of social science in an effort to illuminate the most important and most valued facets of the human mind.

# INFLUENCES

As one with broad training in the social sciences and humanities, I had the good fortune to study and learn—directly or indirectly—from some of the leading thinkers of the 20th century. Upon graduation from college, I worked on an educational research project with Jerome Bruner, one of the founders of cognitive science. And through Bruner, I was introduced to the thinker who had the most influence on my own thinking during the ensuing decades—Jean Piaget, the great scientist of human cognitive development.

Having encountered the works of Piaget and Bruner, I became deeply immersed in their writings. Following in their footsteps, I decided that I wanted to undertake doctoral studies in the area of developmental psychology, with a focus on cognition. During that year of reading, I was also attracted to the ethnography and the theorizing of French anthropologist Claude Lévi-Strauss. I also had the treasured opportunity to listen to (and—as a fan—to shake the hands of) both scholars. As an American immersed in empiricist methods and theorizing, I struggled to understand the common structuralist themes in their writings—as it were, the Cartesian faith that the mind should be construed as a separate domain of experience that can be studied and explicated in scholarly ways.

In what seems to have been my first published article, I linked the thinking of Piaget and Lévi-Strauss in an essay for the broad journal *Social Research*. I sent a draft of the article to both men and was astounded—and thrilled—to receive letters from both of them—incidentally, on the same date, April 10, 1970. What an experience of imprinting that was!

With this unexpected signal of encouragement, I decided to write a book on their individual and joint work, introducing American audiences to the still-mysterious continental field of structuralism. It was titled *The Quest for Mind: Piaget, Lévi-Strauss, and the Structuralist Movement.*

Around that time, I had another encounter that proved to be transformative. During an introductory seminar for doctoral students in

3

developmental psychology, I learned that Nelson Goodman, an eminent philosopher at nearby Brandeis University, was starting a research project in the arts at the Harvard Graduate School of Education. And he was looking for a research assistant.

From my childhood, I was a serious pianist (in fact, I had taught piano informally for a decade), and during my college and graduate school years, I had expanded my interest across the arts more broadly. Though I admired Piaget's view of cognitive development, I could not help but noticing that he always considered, as the apogee of human thought, the mind of the scientist/mathematician/logician.

What would it mean, I pondered, if one were to think of human cognitive development as culminating not in the sciences, but rather in "participation in the arts"?

The opportunity to work with Nelson Goodman and others, in a research project focused on the arts (with the improbable name of "Project Zero"), turned out to be a once-in-a-lifetime stroke of good fortune. To begin with, Goodman was an outstanding thinker and mentor, and he taught me a great deal (given his extremely critical mind, I was relieved to be his quasi-student, rather than one of his doctoral advisees). But equally important, working with Goodman and with a younger group of researchers, I was able to lay out an audacious research project.

That grand (perhaps grandiose) project had two branches:

(1) An empirical effort to examine the development, in children, of capacities that are crucial to artistry. Once one elects to carry out research, one needs a manageable starting point. Accordingly, in the visual arts, I began with the capacity to recognize artistic style; and, in the case of the literary arts, I began with the capacity to understand metaphor.

To give a flavor of this research, I reproduce the abstracts of my first empirical publications—and because journal articles are both dense and somewhat technical, I seek to convey methods and findings in a few words, "in plain English." Thereafter, I provide a summary article about what I had learned more generally about the development of sensitivity to artistic style (see Essay 9). And similarly, focusing on the literary arts, with colleagues, I carried out a set of studies of children's understanding of metaphor. Again, after presenting some findings, I try to put together the picture of metaphoric development.

(2) A conceptual effort to think about human development in terms of the chief *roles* featured in the arts, as a creator, critic, and consumer—and the chief

psychological *systems*—making, perceiving, and feeling. This effort also had two strands: (1) I presented an overview of what it means to traffic in various ways with artistic symbols, and (2) I boldly (or naively!) conceptualized the origins of artistic ways of knowing in the ways that infants first relate to the world. In this section, rather than supply an abstract (the standard summary in scientific journals), I reproduce crucial passages from the text. And this re-orientation led in turn to a conceptualization of what developmental psychology might be like in the post-Piagetian era.

# Jean Piaget

## The Psychologist as Renaissance Man

Enter our young hero, Sebastian. Faced with an adolescent religious crisis, he resolves the issue by adopting an "unshakeable faith" in science. Inspired by his studies in biology, he dreams of a "course synthesizing the sciences of life," granting a privileged niche to psychology and to the theory of knowledge.

Now, for "Sebastian" read "Jean Piaget," the Swiss psychologist and student of child development His novel *Recherché* (*Exploration*) is a personal journal thinly disguised as fiction. It is, perhaps uniquely, Piaget on Piaget: a man who, as a teenager, was a precocious biologist known for his work on mollusks; a man whose natural inclination was to synthesize and who, in his roman à clef, explored for the first time a biological explanation of mental processes.

Ultimately, he considered himself a "genetic epistemologist," searching for the intuitions fundamental to the basic categories of scientific knowledge. Examining numbers, for example, he probed the origins of numerical estimation and classification in the behaviors of the toddler and attempted to reconcile his findings with insights drawn from the philosophy of numbers and the history of mathematics. But Piaget is better remembered for his work in developmental psychology, and more easily assessed in terms of his theories, experimental techniques, and terminology that have and have not survived.

To the public, Piaget was a child psychologist who illuminated the ways in which children's thinking differs from that of adults. With Sigmund Freud and B. F. Skinner, he stood among the most influential figures in 20th-century psychology. If his difficult writing style and lack of concern with the emotional side of human nature made him less of a household word, his position within the mainstream of contemporary psychology is perhaps more secure. And his focus on cognition seems more attuned to the present concerns of psychology than Skinner's concern with overt behavior or Freud's focus on motivation, personality, and the unconscious.

In his capacity to perceive new and important phenomena through the observation of children, Piaget remains without peer. By simply asking children questions and analyzing their errors, exhibiting respect and empathy for his subjects, he entered into children's views of the world. Many scientists were initially skeptical of the results. Many more, however, came to substantiate what Piaget revealed.

He found, for example, that preschool children exhibit "nonconservation"—an inability to believe that matter can change its perceived form and still remain the same in quantity, that a given amount of water poured into containers of different shapes and sizes remains that same amount of water. In exploring childhood egocentrism, Piaget observed that up to a certain age, youngsters cannot conceive how the world looks to others. In infants, he learned that very young children act as if objects fail to exist when they are no longer in sight. Such issues form the mainstay of research in cognitive development today.

To be sure, it is possible to discover the roots of these behaviors in children younger than those to which Piaget ascribed them. But the phenomena still obtain. Psychologists observe and accept them as they develop new theories and devise fresh experiments. Parents themselves, albeit unconsciously, note their children's conceptual limits, and so temper their own actions.

Above all, Piaget's model of the child as an active problem-solver—a hypothesis-testing scientist in shorts—has carried the day. With this image, Piaget not only put the serious study of the child on the scientific map, but moved the child's cognitive powers to the forefront, where they have remained. This sense of what the child is like suffuses the writings of even Piaget's harshest critics. In the manner permitted only the most revolutionary scientists, he changed the way in which future researchers would, and today will, undertake their studies.

Of the terminology Piaget introduced and the concepts he fostered, much has survived but is under constant attack. The biological metaphor he cultivated has been increasingly replaced by the metaphor of the computer; his careful descriptions of behaviors have been supplanted by circuitry detailing the mind's second-by-second "information processing." His discrete, age-related stages of childhood have been blurred. Cognitive development is now seen as a series of smoother transitions.

Piaget's lifelong work—his attempt to explain the cause of intellectual change as a biological tendency to resolve succeeding intellectual challenges in order to return to a conflict-free cognitive state—has had comparatively little influence. People sympathize with this view, but they find the mechanisms he proposed difficult to understand and to test.

Those aspects of a child's life he underplayed—personality, social interactions, artistic gifts—have become rallying posts for contemporary students of child development, if only because Piaget had not already addressed these major issues with the same brilliance that he displayed toward problem-solving. Ironically, then, by what he ignored, as much as by what he illuminated, Piaget set the research agenda for the field he brought to life. His contribution continues to dominate the texts of child study; in more than a few cases, Piaget *is* the text.

## THE PHILOSOPHER'S SHADOW

In an age of fragmented specialization, Piaget was a Renaissance man. Nowhere is this more poignantly evident than in *Logic and Scientific Knowledge,* an encyclopedia planned, edited, largely written by him, and published in 1967, which surveys all the sciences from the perspective of genetic epistemology.

In preparation, Piaget met with experts from every field of knowledge. Rising early each morning, he stayed awake evenings tutoring himself in the discipline at hand, mastering at least its basic conceptual issues. It was Piaget's way of bridging his own science with every other. Indeed, it was his way of pursuing his own religion—the passion for truth, the search for the totality of knowledge.

As an encyclopedia, this work has had little impact. Produced by one man, such a herculean effort must be flawed. Though adept at the study of snails or the children of Geneva, not even Piaget could become sufficiently familiar with other sciences, sympathetic to their histories and cultural backgrounds or sophisticated in their philosophies.

Nonetheless, the book demonstrates that Piaget was as much *philosophe* as psychologist. Almost alone among his contemporaries in psychology, he took the classic epistemological issues seriously, leaving modern philosophers to take Piaget seriously. Indeed, haunting the encyclopedia's 1,250-odd pages is the sense of wonder and exploration that give rise to knowledge. Until the end, then, it seems that the child and the adolescent in Piaget were never stilled. Like Sigmund Freud, he was a passionate speculator and integrator who sought to suppress his speculative nature. Fortunately, he never wholly succeeded.

# Jerome Seymour Bruner
## Cognitive Psychology Enters the Educational Arena

As a student of narrative, Jerome (Jerry) Seymour Bruner knew that one can always tell many stories about an individual person, event, life. Indeed, at the start of his own autobiography, Jerry Bruner wrote "I can find little in [my childhood] that would lead anybody to predict that I would become an intellectual or an academic, even less a psychologist"(Bruner, 1983, p. 4). And yet, it is appropriate—perhaps essential—to begin this appreciation with the fact that Jerry Bruner was born blind. Only at age 2, after two successful cataract operations—Jerry spoke of "good luck and progress in ophthalmology"—could Jerry see. For the rest of his lengthy and event-filled life, he wore memorably thick corrective lenses. And when he was not peering directly at you—be you an audience of one or of thousands—he would grasp his glasses firmly in his palm and punctuate his fluent speech with dramatic gestures.

As a younger child of an affluent Jewish family, living in the suburbs of New York City, Jerry was active, playful, and fun loving—not particularly intellectual or scholarly. His sister Alice wondered why he was always asking questions—Jerry later quipped that he was "trying out hypotheses."

Freud said that the death of his father is the most important event in a man's life. Whether or not cognizant of this psychoanalytic assumption, Bruner seldom referred to his mother; he devoted much more space in his autobiography and much more time in conversation to commemorating his father: "Everything changed, collapsed, after my father died when I was twelve, or so it seemed to me" (Bruner, 1983, p. 5). And indeed, as he passed through adolescence and into early adulthood, Bruner became a much more serious student, a budding scholar, a wide-ranging intellectual. He rapidly negotiated undergraduate life at Duke, became a doctoral student in psychology at Harvard, and received his PhD in 1941, just before the outbreak of World War II. I think that the early death of his father, a decade after the miraculous eye surgery, may have conferred on Bruner ambition, drive, perhaps even a sense of destiny that he might otherwise have lacked.

Over the succeeding 7 decades, Bruner traversed an intellectual landscape as wide as that of anyone in our time. Indeed, while other estimable scholars were writing articles or books in one field or subfield, Bruner swept across departments, even divisions of entire universities, and, extending beyond scholarship, devoted considerable energy as well to areas of practice. Fortunately, in addition to his own lively and trenchant autobiography, several collections of Bruner's writings, as well as a number of biographies and Festschrifts, document my characterization.

When Bruner entered the field of psychology in the late 1930s, it was dominated by the study of sensation (the gateway to perception) and behaviorism (an attempt to explain all animal and human action on the basis of reward and punishment). While he paid his dues in these two traditions, Bruner was never comfortable with a reductionist stance toward human thought and behavior. Following service in World War II, he burst into the headlines—both within psychology and in the popular press—with clever experimental studies carried out in the mid-1940s—studies that were soon dubbed "the new look in perception."

According to this new look, human beings did not simply perceive and then, in an objective manner, report what we were seeing. Rather, perception (whether by sight or another sensory organ) was more trenchantly described as a process of hypothesis formation and confirmation or disconfirmation—we don't know what we see, we see what we know. Dramatically, Bruner and colleagues showed that coins look bigger to impoverished children than to those from affluent families; that it took longer to make sense of anomalous arrays (like a black heart in a deck of cards) than predictable ones; that our guesses about what we perceive are strongly influenced by what we saw before and how likely the alternatives are.

Bruner's introduction of the "new look" forecast well the course that we would follow for the rest of his career. He would create or sense a possible new direction for study; carry out a few pivotal studies; confer a label or name on what he had done; and then move on to another, perhaps entirely unpredictable next challenge, next goal, next world to conquer. For some, including myself, this was an admirable way to proceed; but for many others, this arc signaled an impatience, an unwillingness to dig deeper, a failure to deal with inconsistencies or complexities. As Jerry himself confessed, he was definitely a "fox" who knew many little things, rather than a "hedgehog," who knows one big thing.

Broadly speaking, one can think of Jerry's scholarly career as having three broad themes, each spanning his entire life, but each occupying the forefront for roughly 2 to 3 decades. For the first period of scholarship—roughly 1940s through mid-1960s—Jerry was primarily an experimental psychologist. Following upon his work in social psychology and perceptual psychology, Jerry embarked on the track for which he is best known—opening up the field of cognitive psychology (which is now often collapsed with cognitive science—see Essay 21).

Encouraged and influenced by physicist J. Robert Oppenheimer—the mentor to whom he most often referred—Jerry plunged directly into the mind—the territory explicitly proscribed by most psychologists of the era. In an important work, *A Study of Thinking* (1956), Bruner and colleagues Jacqueline Goodnow and George Austin described how humans go about solving problems—the hypotheses they generate, the solutions they consider, the syntheses that they effect. Shortly thereafter, with his close Harvard colleague George A. Miller, Bruner launched a "Center for Cognitive Studies," in which the full range of scholarly disciplines (from philosophy and linguistics to anthropology and computer science) was brought to bear on how human beings use their minds to the fullest. Nearly every major contemporary actual or aspiring cognitivist passed through the Center, and it was also the chief breeding ground for future leaders of cognitive science.

Bruner's last sustained work in experimental psychology entailed an exploration of the cognitive capacities in young children. He proposed a sequence of "modes of mental representation" in early childhood—from enactive (action) through iconic (perceptual) to symbolic (arbitrary/conventional) ways of representing the world. He also studied the role of nurturing adults and of the broader cultural milieu in the rapid emergence of children's language.

Never one to duck an intellectual skirmish, Bruner's views on representation signaled a battle with the analyses put forth by the noted Swiss psychologist Jean Piaget, and his view of language were an open challenge to the relatively a-cultural and a-psychological views on language put forth by Bruner's sometime colleague Noam Chomsky. Also and prophetically, Bruner expressed his displeasure with the "mainstream of psychology" in a blistering attack published in the (London) *Times Literary Supplement* (Bruner, 1976). Perhaps without quite realizing it, Bruner was in effect relinquishing his membership card in the many psychological societies in which he had once played a leading role.

In September 1959, Bruner convened a conference on education. The meeting brought together an outstanding array of scholars and practitioners from the whole swathe of education. Bruner led the conference masterfully and then, the following year, published a summary of the proceedings, *The Process of Education*. Sharply critical of mainstream approaches to learning, though in a softer "convening voice," Bruner put forth a far more progressive view of education (young people can and should be exposed to cutting-edge ideas across the disciplinary terrain) and a much more constructive view of human cognition (children can ask an apt question, construct knowledge, and revisit issues and concepts at ever-higher levels of sophistication). The slim book also contained the single sentence for which Bruner became best known (and which was most often a subject of debate): "We begin with the hypothesis that any subject can be taught effectively in some intellectually honest form to any child at any stage of development."

For more on Bruner's important work in education, see my collection of papers in *The Essential Howard Gardner on Education*, Essay 1.

It was through his interests in education that I came to know Jerry Bruner. As a 21-year-old college graduate, working at the Underwood School in Newton, Massachusetts, I joined an instructional research team that was developing a curriculum in social studies for middle school students.

Both at the Center for Cognitive Studies and in his work on the curriculum called *Man: A Course of Study* (Bruner, 1965), I was inspired by Bruner's magnetic style of leading; he catalyzed a group of individuals of different ages, backgrounds, and scholarly interests to create a brilliant curricular achievement. When I later assumed leadership roles, I drew on Bruner as a role model.

As suggested, the 1970s marked a sharp break in Bruner's life. He moved to Oxford (perhaps the only professor ever to sail across the ocean in his own boat to take up his new position) to become the Watts Professor of Experimental Psychology; he left the battles of American education; and with his epochal piece in the *TLS*, he in effect abandoned traditional psychological study. When in 1980 he returned to the United States (again as a skipper), he soon moved to New York and embarked on a new intellectual journey.

In this next period, Bruner turned increasingly to the arts and humanities. He proposed that human beings are capable of two quite different kinds of discourse—one more logical and rational, the other more narrative, discursive, and even poetic. Bruner's sympathies were clearly drifting toward this imaginative, this "left hand" way of knowing. Indeed, Bruner explicitly rejected the notion that mental processes could be adequately explained by individual psychology (let alone by study of the brain); to grasp the human mind in its intriguing complexity, one had to understand the surrounding culture, the influence of others, relevant historical and contextual factors, the inevitable conversational and dialogue nature of experience.

Bruner's final years were rich in many unexpected ways. He became a university professor at New York University and, until age 97(!), taught courses on "Culture and the Law," "Vengeance," "Lawyering Theory," and "Narrative and the Law." With Anthony Amsterdam, he published a path-breaking volume, *Minding the Law* (2000), in which the authors illustrated how cognitive, linguistic, and cultural processes can affect the imposition and interpretation of legal processes.

In 1995, Bruner made his first trip to Reggio Emilia, a small northern Italian city known for its remarkable schools for young children. Bruner not only became a careful student—in effect, a visiting teacher—at these schools; he formed personal and professional friendships there; he spent a month there each summer and eventually became an honorary citizen of the community. (For a review of Reggio Emilia, please see *The Essential Howard Gardner on Education*, Essay 3.)

Bruner was an intellectual polymath, which may have made him seem (to some) a gadfly. As he once commented,

> It must have seemed to some of my friends (and critics) that I was jumping fields, although I must confess that it never occurred to me that I was doing anything except follow my nose where it led me . . . Why not simply study what you want to find out letting it take you where it may? . . . . it has not always led me to where I wanted to be nor even to where I thought I would get to. (Bruner, 1983, p. 277)

As a fluent and trenchant writer and a charismatic speaker and commentator, Bruner influenced countless readers—sometimes profoundly. In the late 1980s

I found myself at a conference in Paris with educational scholars from all over the world, almost none of whom I had previously known. A group of us went out to dinner and someone asked, "How did you get interested in education?" An astonishing half of the persons seated at the table said that a major influence had been their reading of *The Process of Education* (1960)!

At his 100th birthday, surrounded by his family and a few close friends, Jerry spoke informally and then, upon request, he recited some verses from T. S. Eliot's "The Love Song of J. Alfred Prufrock." That poem contains the verse "I grow old . . . I grow old . . . ." Amazingly, Jerry Bruner never grew old! As I had written a decade before, "At ninety, he is still the youngest person in the class." No one knows the secret of living vibrantly to a venerable age, but one part of Jerry Bruner's secret is clear: As all who knew him well would attest, he "always looked ahead."

## REFERENCES

Amsterdam, A. G., & Bruner, J. S. (2000). *Minding the law*. Harvard University Press.

Bruner, J. S. (1960). *The process of education*. Harvard University Press.

Bruner, J. S. (1965). *Man: A course of study*. Occasional Paper No. 3. Report for the National Science Foundation.

Bruner, J. S. (1976). Psychology and the image of mind. *Times Literary Supplement*.

Bruner, J. S. (1983). *In search of mind: Essays in autobiography*. Harper & Row.

Bruner, J. S., Goodnow, J., & and Austin, G. (1956). *A study of thinking*. Wiley.

# Project Zero
## Nelson Goodman's Legacy in Arts Education

In the late 1960s and early 1970s, Nelson Goodman founded and directed Project Zero, a research group housed at the Harvard Graduate School of Education. Under Goodman's leadership, Project Zero focused on the nature of artistic knowledge and the ways in which artistic skills and understanding can be enhanced through well-designed programs in schools and museums. Decades later, Project Zero remains an active research center; its current brief is broader, extending well beyond the arts, and involving affiliations with a range of educational institutions. Still, Goodman's original mission—basic research on artistic knowledge and practice—remains a defining feature of Project Zero today.

From one point of view, Nelson Goodman was a most unlikely head of an empirically oriented research project in the area of arts education. Until the mid-1960s, he was a philosopher's philosopher, carrying out fundamental, theoretically oriented investigations of basic issues in epistemology. He had rarely written about the arts. He had had essentially no contact with education below the collegiate level, let alone with those who teach music, visual arts, or drama to the nation's children. And to be honest about it, Nelson Goodman did not have much interest in children or in developmental psychology. He quipped that developmental psychology reduced to the banality that "kids get smarter as they get older." Most of his attention and affection were lavished on his closest colleagues and students, his pets, and the many fascinating works of art that he owned.

As this last clause intimates, from another point of view, Nelson Goodman was a likely head of a project that investigated the basis of artistry. During his undergraduate years at Harvard College (Class of 1928), Goodman had studied with, and been deeply influenced by, Paul Sachs, the associate director of the Fogg Art Museum. Directly following his graduation in 1928, Goodman had for 15 years run an art gallery in Boston. At that time, he had begun to collect art objects, ultimately accumulating collections of such diversity that few appreciated the full ambit of his taste. During World War II Goodman had served as a psychologist in the armed forces and had become fascinated by questions of intelligence, perception, and cognition. Goodman was an inveterate attender of performances across the range of art forms, showing both a discerning eye and ear and a distinct taste for the unusual and the exotic. His wife, Katharine

Sturgis, was a visual artist of some renown; separately and together, they provided material and psychological support for artists and arts projects that captured their fancy.

One day I heard that Goodman—whom I knew slightly by reputation—was looking for research assistants to staff a new project in the arts and arts education. I remember my first meeting in an office at nearby Brandeis University. Goodman was formidable, intimidating. He said to me, "Do you read philosophy?" I told him that I was reading the phenomenologist Maurice Merleau-Ponty. Goodman winced noticeably. I then said that I had also been influenced by Susanne Langer, who had written an important study of human symbol use called *Philosophy in a New Key* (1942). Goodman relaxed noticeably: "That's a different story."

Unlike some others in the Project Zero ambit, I always liked Goodman. To be sure, he was tough, cranky, opinionated, bossy. But he also had a softer, more gentle side, particularly when it came to the arts. Memorably, he said, "Ask not what the arts can do for you; ask what *you* can do for the arts." He was a true amateur, a true lover of the arts. He collected art and had a terrific eye. Goodman was also an *impresario manqué*. He liked staging arts events—dances, theater, musicales, even multimedia shows. He produced and directed at least a dozen arts and lecture performances, including ones involving quite famous artists of the time.

The one that made the deepest impression on me was a lecture performance about Arthur Miller's play, *Death of a Salesman* (1949). The audience watched several different live and videotaped/filmed performances of the same scene from the play, and then George Hamlin, the director of the Loeb Drama Center at Harvard, discussed the various moves we had observed. (Christopher Reeve, then a teenager!, was in one of the productions.) I could mention many other Goodman productions. He brought the arts to the Harvard Ed School in a way that they had never been brought before—or since.

Still, it is probable that Project Zero would never have been launched had it not been for the concatenation of three factors.

First of all, in the 1960s, Nelson Goodman completed a lengthy labor of love: his book *Languages of Art* (1968), in which he put forth an approach to the arts grounded in the study of different symbols, symbol systems, and modes of symbolic functioning. While the book was heavily theoretical, Goodman also drew on the findings of psychology, linguistics, and other empirically oriented disciplines. And in a pregnant final passage, he speculated:

> We hear a good deal about how the aptitudes and training needed for the arts and for the sciences contrast or even conflict with one another. Earnest and elaborate efforts to devise and test means of finding and fostering aesthetic abilities are always being initiated. But none of this talk or these trials can come to much without an adequate conceptual framework for designing crucial experiments and interpreting

their results. Once the arts and sciences are seen to involve working with—inventing, applying, reading, transforming, manipulating—symbol systems that agree and differ in certain specific ways, we can perhaps undertake pointed psychological investigation of how the pertinent skills inhibit or enhance one another, and the outcome might well call for changes in educational technology . . . The time has come in this field for the false truism and the plangent platitude to give way to the elementary experiment and the hesitant hypothesis. (Goodman, 1968, p. 265)

A second factor that made possible the creation of Project Zero was the educational atmosphere in the United States of the 1960s. Following the unexpected launch of the satellite Sputnik by the Soviet Union in 1957, the U.S. government devoted unprecedented sums of money to the improvement of scientific, mathematical, and technological education, particularly at the elementary and secondary levels. Little or no attention was being directed to education in the arts and humanities. By the mid-1960s, both private and public granting agencies were struck by this asymmetry and determined to correct it, at least a bit.

The final spur to Project Zero was a particular cast of characters located in a specific geographic and temporal region. At the Harvard Graduate School of Education, Dean Theodore Sizer (himself the son of a professor of fine arts) was looking for ways to foster research and pedagogy in the arts. A key member of the education school faculty was Israel Scheffler, a distinguished philosopher with an interest in the arts and symbolization and, not coincidentally, Nelson Goodman's one-time student and lifelong friend.

Sometime in 1967, Project Zero was christened and launched. The name reflected the founder's estimate of the state of *firm knowledge* about arts education. Goodman moved his office (and his concept) from the Brandeis campus to Harvard Square. Equipped with a little budget and a lot of lively ideas, Goodman launched an unprecedented research project that has lasted, indeed prospered, until today (i.e., 2024). In a 1988 essay, Goodman recalled:

When Project Zero started twenty years ago with a staff consisting of one philosopher, two psychologists part time, and some volunteer associates, it had no fixed program and no firm doctrines but only a profound conviction of the importance of the arts, and a loose collection of attitudes, hunches, problems, objectives, and ideas for exploration. We viewed the arts not as mere entertainment, but like the sciences, as ways of understanding and even of constructing our environments, and thus looked upon arts education as a requisite and integrated component of the entire educational process. We found at once that we had to begin almost at zero, with basic theoretical studies into the nature of art and of education and a critical scrutiny of elementary concepts and prevalent assumptions and questions. (1988, p. 1)

As one of the two part-time (and largely uncompensated!) psychologists to whom Goodman refers, I have had the privilege of being associated with Project Zero since its inception. The early years of the Project were very exciting. Goodman was intellectually charismatic. He had developed a rigorous approach to the analysis of artistic work and process, and he was eager to see it applied: to the analysis of aesthetic concepts (like style, rhythm, metaphor); to the launching of psychological experiments (on how we perceive perspective; how we group patterns rhythmically; whether children can perceive artistic style); to the education of teachers (through the staging of lecture performances by well-known artists representing different media and genres); and even, as I was soon to become the beneficiary, to the brain bases of artistic knowledge and experience.

The aforementioned motley crew would regularly gather around a lengthy table, discuss these issues on a theoretical level, design and critique psychological experiments, listen to (and respectfully challenge) learned speakers, and make sure that the plans were in place for the forthcoming lecture performances. The table is easier to recall than the building, because Project Zero was like an orphan during its first years.

Not really being part of the school's "line item" educational program, we were tossed about annually from one building and from one part of the campus to another. Nelson Goodman locked horns with changing administrations, and did so with tenacity and gusto. I recall one particularly dreary moment when, after months of bickering, it was still not clear whether we had secured a rug for the office. Asked by his friend Paul Kolers whether this experience had tested Nelson's faith in human nature, Goodman growled, "No, it confirms it!"

Project Zero accomplished a great deal during the 5 years that Nelson Goodman stood at the helm. The Goodman era showed that a dogged group of scholars and researchers could bury (or at least bracket) their disciplinary differences and carry forward solid philosophical, psychological, and educational work in the area of arts education. Since we were scholars, the products were mostly ideas and writings. (This characterization also applied to the Center for Cognitive Studies; see Essay 2.)

These writings by early Project Zero personnel are scattered in many places—one spot where their cumulative effect can be gleaned is the 1977 book *The Arts and Cognition,* edited by David Perkins and Barbara Leondar.

But probably the most precious legacy of the Goodman era is a 100-page unpublished "final report" for the Office of Education, supervised in 1972 by Nelson Goodman. Entitled *Basic Abilities Required for Understanding and Creation in the Arts,* this monograph sets forth the basic assumptions of the Project, surveys the major studies carried out and results achieved to that point, and attempts to evaluate whether "Project Zero has moved toward one."

The report begins with a simple assertion of the purpose of the Project: "the advancement of the arts through improved education of artists, audiences, and management." It laments the primitive nature of work in the area of arts education; it chronicles the many obstacles of funding and attitude that lie in the way of a cognitive approach to the arts; but it declares with a touch of pride, "in the course of our study to date some promising hypotheses have gained support" (Goodman et al., 1972, p. 1). It goes on to review the major philosophical issues and controversies in aesthetics; clarification of concepts in the various art forms; studies of perception and cognition in children and adults; how one might evaluate artistic programs; and various "field" initiatives, including site visits to institutions of arts education, sponsoring of lecture performances, the initiation of a training program in arts management, and the launching and evaluation of a course on Project Zero.

An examination of this list in the 21 century might suggest that Project Zero was situated in the mainstream of research in the late 1960s. Nothing could be further from the truth. The ideas of cognitive psychology and psychologically informed philosophy were still relatively new in the academy. They were newer still in the area of education. And they were considered virtual blasphemy in arts education.

Whatever artists and arts educators may have believed in their souls (an entity that Nelson Goodman would not have endorsed!), talk of arts focused very much on emotions, spirit, mystery, the ineffable and unanalyzable. An effort to demystify the arts, to construe them as involving the same *kinds* of skills and capacities as are involved in other domains and other disciplines, was incendiary. And an effort headed by an epistemologist, and staffed primarily by experimentally oriented social scientists, was deeply suspect. As Goodman later recalled, "In the early years of the Project, the very idea of the arts as cognitive and systematic research into arts education met with widespread and virulent hostility" (Goodman, 1988, p. 2). We once received a rejection of a grant application containing the admonition: "If this research were carried out, the arts would be destroyed." (So much for *ars longa!*) And shortly after Nelson Goodman and Ted Sizer retired (both in 1972), the administration of the Harvard Graduate School of Education literally tried to kill Project Zero by forbidding us to apply for further funding. Only a heroic act of diplomacy by Israel Scheffler saved our still-fledgling organization.

Elsewhere I have presented the history of Project Zero in the post-Goodman era (Gardner, 2013). As of 2024, at any one time, there are between 50 and 100 active researchers involved with the Project, working on 20 to 30 different projects and offshoots thereof (see pz.harvard.edu).

It is fair to ask what Nelson Goodman thought of the Project Zero of the 1970s, 1980s, and 1990s. On one hand, Goodman deplored Project Zero's ever-expanding scope of inquiry, particularly when it deviated from the arts. Recall that this highly disciplined, almost ascetic man once quipped, "Ask not what the arts can do for you; ask what *you* can do for the arts." He never liked my theory of multiple intelligences, for its conceptual basis eluded him. (He once threatened to send the *New Yorker* a satire called "The Seven Stupidities.") He was suspicious of Project Zero's efforts to improve education directly. As he once wrote:

Our task is to provide analysis and information that may help in clarifying objectives and concepts and questions, in avoiding some pitfalls, in recognizing obstacles and perceiving opportunities . . . When Project Zero turns to writing prescriptions and instruction books, when it becomes Project-How-To, it will have passed on to an unjust reward. (Goodman, 1988, p. 1)

On the other hand, he kept in regular touch with activities at Project Zero and happily joined us in celebrating the 15th and the 25th anniversaries. I think that he would agree that we have moved from Zero to One, and perhaps Beyond.

## WHAT OF GOODMAN'S LEGACY TO PROJECT ZERO?

To begin with, Nelson Goodman provided us with a bountiful set of concepts and issues, whose exploration can continue to occupy generations of students and scholars across a range of disciplines. Second of all, we were inspired by an impressively high set of standards of thinking, critiquing, and writing; one did not submit shoddy work to Goodman, and his own stern superego has been internalized by many of us . . . though, no doubt, insufficiently, he would have reminded us. We absorbed a fascination with nonobvious problems and with seemingly oxymoronic notions: How can a Project be Zero? How can intelligence be multiple? What can the brain possibly tell us about the classification of symbol systems? How can understanding be a performance? We received a respect for ivory tower analysis, but at the same time were impressed by the need for grounding in real works of art, and in direct contact with artists, those who manage arts programs, and the audiences without which no artistic genre can survive. Indeed, one of Nelson Goodman's chief (though largely unsung) contributions to Harvard and the Boston community was his willingness throughout his last years to sponsor innovative arts programs that exemplified his faith in the cognitive aspects of the arts.

Speaking personally and on a first-name basis, I took from Nelson an irreverent but not unhopeful view of the human condition. As I've noted, Nelson could be tough on himself and on others (I think that he saw himself as a "man of the world"); but he had a wonderful, and touchingly naïve, faith that the cause of the arts and cognition will advance. Once we were sitting around, trying to figure out how to help someone become more creative. One person suggested the provision of interesting problems; the other suggested the technique of brainstorming. Nelson shook his head and said, "Create obstacles, and make sure that they are productive ones." I am reminded of Freud's words at the end of *The Future of an Illusion*: "The voice of the intellect is a soft one, but it does not rest until it has gained a hearing" (1927/1950, pp. 96–97). Nelson Goodman's work has advanced this cause in those spheres of knowledge that had proved most resistant: Nelson's restless voice, words, and example represent his enduring contribution to understanding and to practice in the arts.

## REFERENCES

Freud, S. (1950). *The future of an illusion*. Hogarth Press. (Original work published 1927)

Gardner, H. (2013). Harvard Project Zero: A personal history. https://pz.harvard.edu/sites/default/files/pz-history-9-10-13.pdf

Goodman, N. (1968). *Languages of art*. Bobbs-Merrill

Goodman, N. (1988). Aims and claims. *Journal of Aesthetic Education, 21*(1), 1–2.

Goodman, N., Perkins, D., & Gardner, H. (1972). *Basic abilities required for understanding and creation in the arts*. Final Report for the U.S. Office of Education. Harvard University.

Langer, S. K. (1942). *Philosophy in a new key: A study in the symbolism of reason, rite and art*. Harvard University Press.

Perkins, D., & Leondar, B. (1977). *The arts and cognition*. Johns Hopkins University Press.

# Norman Geschwind as a Creative Scientist

I was fortunate to have been a colleague and friend of Norman Geschwind, one of the indisputably creative scientists of our era. I learned much about creativity from observing Norman, and some of what I have learned has been useful to me as I have thought more broadly about the parameters of creativity (see Essays 23 and 24). In these notes, I attempt to capture this interplay in my own thinking about Norman Geschwind as a creative scientist—as well as a few thoughts about the more general phenomena of creativity.

To begin with, Norman Geschwind was a man of great raw *intellectual power*. No matter what one's definition of intelligence, or intelligences, it was clear that Norman (or Norm, as I was privileged to call him) stood out in terms of the power and penetration of his mind. His verbal skills were legendary—speaking, writing, readily mastering languages, reading at a prodigious rate, exhibiting impressive sensitivity to how language works and how it breaks down. Norman had equally powerful gifts in the area of logical-mathematical thinking; he began as a mathematics major in college and never lost the habit of thinking about issues in a logical, rational, and often mathematical way. (Addendum, 2021: I just learned that in college he shifted his major to Social Relations, the same interdisciplinary area that I studied 20 years later!)

Norman was less outstanding in the realms of spatial thinking or in the use of his hands—in this way, he did not resemble many benchtop scientists. But his interests swept across all domains. He confessed to very little competence in music. It is therefore especially notable that he encouraged people like me to study music; that he collaborated with Tedd Judd and me in a study of a composer who became verbally (though not musically) alexic (Judd et al., 1983); and that he frequently attended concerts, including one just hours before his untimely death at the age of 58 in November 1984. Until the very end of his life, clearly Norman wanted to learn about almost everything.

In addition to his polymathic intellect, Norman had a kaleidoscopic capacity to draw on his knowledge. Virtually everything that he knew—and he knew more about more topics than just about anyone—was available at his cerebral fingertips. Any new observation or fact could kindle a startling chain of associations, a new pattern, as that kaleidoscope went famously to work and to play.

I suspect that if one had been able to perform a scan on Geschwind's brain in operation, one would have beheld a uniquely beautiful series of pictures.

Working hand-in-glove with intellect are the *personality structures* of the individual. Norman Geschwind displayed in abundance all the personality traits that are concomitants of great creativity—a driving personality, high degrees of energy, clearly focused attention, perseverance, a willingness to take risks, and a reassuring amount of self-confidence. Those in search of creative role models had no better option than to spend time in the presence of Norman Geschwind. Moreover, Norman had an expansive *network of enterprise*—a large number of individuals and projects that he helped to sustain and that in turn kept him fully occupied and engaged (Gruber, 1981).

But while Norman resembled the textbook case of the creative personality, he also gave us a whole ensemble of bonuses. He was wonderfully funny; he was a pleasure, a treat, a perpetual feast to be around. He was without a trace of vanity or self-importance. One could never tell a person's status simply by observing his or her encounters with Norman from afar; Norman would spend hours with a naïve undergraduate even as he would not hesitate to put a pretentious senior scientist in his place. He set an enviable standard for collegiality, responding to letters and phone calls promptly, and often at admirable length, almost always answering requests favorably, putting himself out for others in ways that were not necessary and that should have filled more of us—the unbridled requesters—with embarrassment. Perhaps it is best to say that Norman made us feel that he was one of us, though, deep down, we knew that he was one of the immortals.

Norman Geschwind was a true scholar and could probably have made contributions in an ensemble of domains, ranging from mathematics to history to linguistics. But his chosen domain was behavioral neurology, and it was there—and in neighboring disciplines—that he was to make his enduring contributions.

Norman's relation to the domain of behavioral neurology was peculiar. He did not invent the field, nor did he pull it in wholly original directions. Rather, he rediscovered a tradition that had existed a century earlier but had been allowed to become dormant. Geschwind revivified the discredited works of earlier scholarly giants—the so-called "diagram makers"—who attempted to locate specific cognitive facilities in specific brain regions or the connection among them. In the days when doctrines of holism and equipotentiality reigned supreme, this was an act of intellectual courage.

Of course, Norman Geschwind did not just direct his undereducated contemporaries to consult dusty superannuated journals. He described new cases that either confirmed the traditional descriptions or that modified them in fresh and illuminating ways. His work on the disconnection syndromes had become classic within a decade of its original publication (Geschwind, 1965, pp. 237–294). His papers on the aphasias, the apraxias, the alexias, temporal lobe epilepsy, and many other neuropsychological syndromes and conditions brought classical typologies in touch with newly observed phenomena and contemporary modes of analysis. He himself was in contact with the outstanding scientists in Russia,

Europe, Asia, and the Americas; and he brought the rest of the community of
brain and behavior in touch with these same authorities.

Nearly everyone who heard Norman Geschwind concurred that he was one
of the great scientific lecturers of the age. He could discuss complex issues in
a straightforward way, bring new excitement to a discussion, and respond to
even the most vexing questions with insight, appropriateness, and timely wit.
Through his effect on others, directly and indirectly, he almost single-handedly
and single-mindedly redirected the standards of what lines of work were important and how such issues could be addressed in an intellectually rigorous way.
Indeed, as happens in such cases, Norman exerted a powerful influence even on
those who staunchly disagreed with his claims.

Under his founding leadership, a number of important research and treatment centers were launched—and I was privileged to be affiliated with them and
to have Norman as a mentor for more than a decade. For years, there was but
a tiny minority of behavioral neurologists and neuropsychologists who had *not*
been trained by Norman Geschwind; and even those who had not been trained
at his scintillating "patient rounds" flocked to his lectures and combed over recently published journals for his writings—and, eventually, for the writings of
his students.

A dimension entailed in the phenomenon of creativity is often overlooked—
the dimension called the *field*. Whatever the talent of the individual, and the state
of readiness of the domain, a creative effort cannot come to be seen as such without the collaboration of those individuals, institutions, award panels, and the
like that render judgments of quality (Csikszentmihalyi, 1988, pp. 325–339).

The field recognized Norman Geschwind's talents at an early age. He won
more than his share of fellowships, residencies, grants, prizes, professorships, and
other forms of recognition.

But I'd like to stress another aspect of the field. Even as Norman helped to
reorient the domain of behavioral neurology and its neighbors, he was crucially
important in helping to shape its concomitant fields. Norman's judgments about
work, his taste with respect to issues worthy of study, his ability to shape discussions and standards were remarkable and, to my way of thinking, wholly
beneficent. Norman Geschwind exerted this effect by his writings, his choice
and development of students, his engagement at scientific meetings, and, perhaps especially, his lectures. And I was a personal beneficiary of this facet of
Norman's life. While he was not a developmental psychologist, he encouraged
me to link findings from neuropsychology and child development; while he was
not oriented toward the arts, he encouraged me to study artistic development
and breakdown under conditions of brain damage; and while he was not a book
writer, he helped me write my book *The Shattered Mind (1975)*, which I was
pleased to dedicate to him.

Norman Geschwind stood at the center of what Margaret Mead once called
a cultural cluster, "whose defining characteristic is at least one irreplaceable
individual, someone with such special gifts of imagination and thought that

without him the cluster would assume an entirely different character—a genius who makes a contribution to evolution not by biological propagation but by the special turn that he is able to give to the course of cultural evolution" (Mead, 1964).

I trust that, in terms of the criteria that I have defined, Norman Geschwind emerges as an exceptional individual, one who embodies creativity of a very high order. His combination of probing intellect, driving yet generous personality, substantive reorientation of behavioral neurology, and shaping of the field of clinical neuropsychology places him centrally in that privileged group of scientists who have made a difference in our time.

## REFERENCES

Csikszentmihalyi, M. (1988). Society, culture, and person: A systems view of creativity. In R. Sternberg (Ed.), *The nature of creativity* (pp. 325–339). Cambridge University Press.

Gardner, H. (1975). *The shattered mind: The person after brain damage*. Knopf: distributed by Random House.

Geschwind, N. (1965). Disconnexion syndromes in animals and men. *Brain, 88,* pp. 237–294.

Gruber, H. (1981). *Darwin on man*. University of Chicago Press.

Judd, T., Gardner, H., & Geschwind, N. (1983). Alexia without agraphia in a composer. *Brain, 106,* 435–457.

Mead, M. (1964). *Continuities in cultural evolution*. Yale University Press.

# EARLY WORK

Having encountered the works of Piaget and Bruner, I became deeply immersed in their writings. I decided that I wanted to undertake a doctoral program in the area of developmental psychology, with a focus on cognition. During that year of reading, I was also attracted to the ethnography and the theorizing of French anthropologist Claude Lévi-Strauss. I also had the treasured opportunity to listen to (and to shake the hands of) both scholars. As an American immersed in empiricist methods and theorizing, I struggled to understand the common structuralist themes in their writings—as it were, the Cartesian faith that the mind was a separate domain of experience that could be studied and explicated

In what may well have been my first published article, I linked the thinking of Piaget and Lévi-Strauss in an essay for the broad journal *Social Research*. I sent a draft of the article to both men and was astounded—and thrilled—to receive letters from both of them—indeed, on the same date April 10, 1970. What an experience of imprinting that was!

With this unexpected signal of encouragement, I decided to write a book on their individual and joint work, introducing American audiences to the still mysterious continental field of structuralism. Published in 1973, it was titled: *The Quest for Mind: Piaget, Lévi-Strauss, and the Structuralist Movement.*

Around that time, I had another encounter which was transformative. During an introductory seminar for doctoral students in developmental psychology, I learned that Nelson Goodman, an eminent philosopher at nearby Brandeis University, was starting a research project in the arts at the Harvard Graduate School of Education. And he was looking for a research assistant (see Essay 3).

From my childhood, I was a serious pianist (in fact, I had taught piano informally for a decade), and during my college and graduate school years, I had expanded my interest across the arts more broadly. Though I admired

Piaget's view of cognitive development, I could not help but notice that he always considered—as the apogee of human thought—*the mind of the scientist/mathematician/logician.*

What would it mean, I pondered, if one were to think of human cognitive development as culminating not in the sciences but rather in "participation in the arts"?

The opportunity to work with Nelson Goodman and others, in a research project focused on the arts (with the improbable name of "Project Zero") turned out to be a once-in-a-lifetime stroke of good fortune (see Essay 3). To begin with, Goodman was an outstanding thinker and mentor and he taught me a great deal (given his extremely critical mind, I was relieved to be his quasi-student, rather than one of his doctoral advisees). But equally important, working with Goodman and with a younger group of researchers, I was able to lay out an audacious research project.

That grand (perhaps grandiose) project had two branches:

(1) An *empirical* effort to examine the development, in children, of capacities that are crucial to artistry. Once one elects to carry out research, one needs a manageable starting point. Accordingly, in the visual arts, I began with the capacity to recognize artistic *style*; and, in the case of the literary arts, I began with the capacity to understand *metaphor.*

To give a flavor of this research, I reproduce brief sections of my first empirical publications (Essays 7 and 8); and because journal articles are both dense and somewhat technical, I describe their findings "in plainer English." Thereafter, I provide a summary article about what I had learned more generally about the development of sensitivity to artistic style (Essay 9). And similarly, focusing on the literary arts (Essays 10 and 11), with colleagues, I carried out a set of studies of children's understanding of metaphor.

(2) A conceptual effort to think about human development in terms of the chief *roles* featured in the arts, as a creator, critic, consumer—and the chief psychological *systems*—making, perceiving, feeling; this effort also had two strands: (a) conceptualization of the origins of artistic ways of knowing in the ways that infants first relate to the world (Essay 6), and (b) an overview of what it means to traffic in various ways with artistic symbols (Essay 11).

Again, rather than an abstract (the standard summary in scientific journal), I reproduce crucial passages from the text. And this reorientation led in turn to a conceptualization of what developmental psychology might be like in the post-Piagetian era (Essay 12).

# Piaget and Lévi-Strauss
## The Quest for Mind

*In addition to my interest in the arts, I also continued to be fascinated by the work of Jean Piaget—and I discerned surprising similarities with the work of another leading Francophonic social scientist, the French anthropologist Claude Lévi-Strauss. In this article, I describe their work and do a classical "compare and contrast." And while it was not the focus of the article, in the concluding paragraphs I attempt to link their scholarly efforts to the understanding of the arts.*

*In my innocence, I—a mere doctoral student—sent a version of this paper to both Piaget and Lévi-Strauss. To my amazement, they both responded promptly—and to my astonishment, on the same day (April 10, 1970). In addition to most of the article, I reproduce those letters (see Figure 5.1 and Figure 5.2). As a tribute to both giants, I always respond promptly—and, I hope, helpfully—to students who write to me.*

*I was able to maintain contact with both scholars, and a few years later, I published my first book for the general reader,* The Quest for Mind: Piaget, Lévi-Strauss, and the Structuralist Movement *(Gardner, 1973).*

I

A method and a puzzle, both of longstanding fascination for those schooled in the French tradition, have recently been joined with great elegance. Jean Piaget and Claude Lévi-Strauss, whose prolific and ponderous works are exerting increasing influence on contemporary social science, have wielded the tools of structural analysis in an effort to dissect the human mind. Neither thinker would willingly acknowledge such Cartesian dualism: Piaget, steeped in biology, stresses that thought is rooted in the biological evolution of the organism; Lévi-Strauss, a self-avowed Marxist, asserts that the products of mind are ultimately reflections of the society's infrastructure. Nevertheless, when viewed from a transatlantic or trans-Channel perspective, both men can be fairly regarded as pursuers of what critics have described as the "Ghost in the Machine."

The similarities in life and style are striking: both born in the shadows of 1900, both precocious students (Piaget issuing papers on malacology while still in his teens); both early converts to Henri Bergson and "inconstant disciples"

## Figure 5.1. Letter from Claude Lévi-Strauss

LABORATOIRE D'ANTHROPOLOGIE SOCIALE

DU COLLÈGE DE FRANCE ET DE L'ÉCOLE PRATIQUE DES HAUTES ÉTUDES

TÉL. 633-76-10 OU
033-61-60 ET 326-26-53 (POSTE 211)

11 PLACE MARCELIN-BERTHELOT
PARIS 5

Paris, April 10, 1970

Dr. Howard Gardner
Harvard University
Department of Social Relations
William James Hall 1257
Cambridge, Massachusetts 02138
U.S.A.


Dear Mr. Gardner,

Many thanks for sending me your two papers which I read with great interest. I fully agree with the parallel you draw between Piaget and I, except that it seems to me that you have overlooked two points. In the first place, Piaget is substantially older than I am and since he started publishing very early he was already read and lectured upon when I was still a student at the University. In the second place, there are in fact two Piagets, the early one with whom I am not in great sympathy and the later one whom, on the contrary, I admire greatly

Concerning your other paper, to be quite frank, the kind of criticism you level at me makes me shrug. I look at myself as a rustic explorer equipped with a woodman's axe to open a path in an unknown land and you reproach me for not having yet drawn a complete map, calculated accurate my bearings and for not having yet landscaped the country ! Forgive me f saying so but it looks as if Lewis and Clark were taken to task for not having designed the plans of General Motors while on their way to Oregon Science is not the work of one man. I may have broken new ground but it will take a great many years and the labour of many individuals to till and make the harvest.

With best regards.

Sincerely yours,

Claude Lévi-Strauss

CLS:eg

**Figure 5.2. Letter from Jean Piaget**

# UNIVERSITÉ DE GENÈVE

SCHOLA GENEVENSIS MDLIX

—

## FACULTÉ DES SCIENCES

—

Genève, le 1o avril 197o

Monsieur Howard GARDNER
Harvard University
Department of Social Relations
William James Hall 1257
CAMBRIDGE , Mass. o2138

Cher Docteur Gardner,

J'ai bien reçu et j'ai lu avec grand
plaisir vos deux belles études sur Levi Strauss et sur moi.
Elles sont remarquables de finesse et de pénétration et je
vous en félicite. C'est un plaisir de lire des gens qui
vous comprennent et cela n'arrive pas tous les jours.

J'aimerais seulement vous signaler à
propos de l'article de Parsons sur mon utilisation de la
logique formelle, que Parsons m'a très mal compris. Le mathé-
maticien et logicien Seymour Papert, qui enseigne au M.I.T.
et que vous connaissez sans doute, avait fait une réponse
pour montrer que ma logique était bien sûr différente du cal-
cul des propositions classiques, mais qu'elle était consis-
tante et répondait bien aux besoins psychologiques voulus.
Malheureusement le British Journal of Psychology a refusé
de publier cette étude de Papert, pour éviter une polémique.
J'ai trouvé cela malhonnête, mais surtout regrettable, parce
que Papert disait des choses nouvelles et importantes.

D'autre part, le logicien belge Apostel
a fait un article dans le même sens, mais qui n'a pas non
plus paru. Par contre d'autres logiciens, et en particulier
Jean-Blaise Grize, ont décidé de rééditer mon Traité de logiqu
en fournissant aux lecteurs des informations suffisantes pour
qu'ils comprennent les points de vue nouveaux qu'il contient.
Je vous signale tout cela pour que vous puissiez renseigner
éventuellement vos lecteurs sur ces points où il y a eu malen-
tendu complet avec les logiciens.

En vous remerciant encore très vivement,
je vous prie, cher Docteur Gardner, de croire à mes sentiment:
les meilleurs.

J. PIAGET

of Émile Durkheim, both impatient with the "illusions" of philosophy (notably existentialism); and both influenced by modern mathematics and enamored of formal elegance. Their scholarly careers have clear parallels as well; the years have seen each man entering upon explored paths, working initially in isolation, examining the relationships among the various disciplines, inspiring much abuse and misunderstanding (except among a small group who "understand well"), embracing a structural approach to the social sciences. Somewhat poignantly, each scholar has looked toward certain not-quite-civilized Frenchmen—little children or grown-up "savages"—to cast a new light on the nature of his own mind. With a list of parallels that could be extended indefinitely, a recent aside by Piaget becomes persuasive: It is "unthinkable" that there have been virtually no attempts to relate to one another these two strands of thought.

## II

Few projects in modern thought have been so sweeping in design and so doggedly pursued as the genetic epistemology of Jean Piaget. Set forth in a philosophical novel of his *Wanderjahre* is Piaget's goal of relating the part to the structured whole at every level and in every domain of biology and knowledge. Over a 50-year period, Piaget launched and sustained a dual assault on this problem: He examined the genesis and the development of a staggering range of human capacities in the growing child, including the child's conceptions of objects, space, time, causality, numbers, morality, and "the world"; he outlined again and again the nature of each scientific discipline, the methods used, the underlying formal models, and the genuine and spurious relationships to other fields. Piaget's procedures have become his hallmark: careful questioning and ingenious observation of a small number of children; extensive reliance on his own intuitions about what the child really knows; and, increasingly in the latter phases of his career, intensive collaboration with scientists and logicians on interdisciplinary issues. The science and the sciences of man become organized in Piaget's synthetic mind and (given his mode of behavior) simultaneously in his writings, of which roughly a thousand pages were published yearly. A brief characterization of his work helps elucidate Piaget's relationship to Lévi-Strauss.

The psychologist and the biologist in Piaget are equally insistent that knowledge proceeds from human action. Any study of knowledge must begin with an identification of the primitive mechanisms by which stimulation is initially processed—the world can only be assimilated at the child's level of development. Though we may be certain that the world contains objects for the child to use and know, the child is himself privy to no such insight; he must literally and painstakingly construct a world of objects. Only following months of looking at, picking up, and searching for objects does the child understand that these objects have an existence apart from his own actions and view, that these objects can be used to accomplish diverse ends. And so it is with the other Kantian categories (e.g., space, time), none of which is present *a priori*, all of which will develop into adult form, given a normal child and sufficient interaction with the environment.

Piaget insists that all knowledge, including that rarified blend produced in his Genevan laboratory, must be constructed through the person's actions, performed either in the world out there or as interiorized actions. Such internal actions or operations, while occurring within the mind, are akin to direct manipulation of material objects. Even when observing, the cognizing child is actually transforming what he sees. For example, only when a child is able to participate implicitly in the pouring (and to anticipate the inverse of this action) does he realize that the amount of water remains constant, irrespective of the dimensions of its container. Similarly, only because he has enacted the role of other persons on numerous occasions does the child grasp that persons seated across the table from him will see a scene differently, and that those seated next to him may hold very different views about the world.

Piaget's genius is shown by his demonstration that such capacities, which everyone had more or less assumed to be present from infancy, only emerge ordinarily at about the age of 7 or 8, following upon intensive experimentation with the world of objects.

According to Piaget, intellectual development does not cease with these "conservations," but continues, at least in Western society, until adolescence. One capable of formal operations can reason not merely about physical occurrences but also with and about linguistic (or mathematical) propositions, referring either to material events or to abstract philosophical issues. Even at such scholastic levels of discourse, however, the thinking is invariably active: The adolescent still performs the actions—adding, reversing, deleting—but now they are performed on verbal propositions, rather than merely objects in the world.

For some, Piaget's contributions consist chiefly of the fascinating insights he has gleaned about the minds of children. Piaget's own enterprise is far more ambitious, however. He has developed models for each of the principal stages of thought: These are statements, expressed in formal logic, that describe the operations of which the youngster is psychologically capable at a given level of cognitive development. Piaget's formalism at first drew the scorn of logicians, who saw little point in formalizing incorrect or incomplete thinking, and who spotted numerous lapses on the part of the amateur logician. It has proved possible to "straighten out" Piaget, however; now an entire Center of Genetic Epistemology is busily applying the formal models to children's reasoning and to scientific thought.

Along with Piaget's contention that thought derives from action comes a heretical view: Language does not play a determining role in intellectual development. Piaget insists that knowledge which stems principally from language is never genuine knowledge, that words obfuscate the child's understanding of a phenomenon; it's the attainment of operations (and a higher level of thought) that determines changes in linguistic capacity, rather than the reverse.

## III

Whereas Piaget consistently questions the importance of language, Lévi-Strauss derives from this human faculty his entire conception of thought. "Language . . . is human reason," he asserts (Lévi-Strauss, 1966, p. 252). To be sure, Lévi-Strauss has a highly specific view of language, one borrowed essentially intact from the Prague school of linguistics and from his mentor, Roman Jakobson. According to this structuralist approach, the analyst must specify meaningful units, note the "distinctive features" of which these units are comprised, and then, rather than treating the units as independent entities, focus upon the relationships among them.

Lévi-Strauss's earliest (and most faithful) application of structural methods was in the anthropologist's heartland—*kinship relations*. The principal challenge: the determination of appropriate units of analysis. For example, in attempting to understand the importance in family relations of the maternal uncle (or avunculate), cultural anthropologists had routinely seized upon individual kin terms or on the relationships within the nuclear family. By making individual terms the units, Lévi-Strauss's predecessors had too literally adhered to the linguistics method, thereby betraying its essence.

Lévi-Strauss's conclusion: The appropriate unit for explanation of the avunculate included the positive or negative attitudes held by eight relatives toward one another. The "elementary structure" that he uncovered was a "global system" in which the relationship (or attitude) between maternal uncle and nephew was to that between brother and sister as the relationship between father and son was to that between husband and wife. Accordingly, if one knew certain attitudes held by relations, it became possible to infer the remaining ones.

Taking his formalism where he found it, Lévi-Strauss attempted to apply algebraic models to kinship structure and to utilize the insights of information and game theories in his subsequent studies of myth and of social organization.

During the 1960s, a period of prodigious, virtually Piagetian output, Lévi-Strauss employed formal procedures and linguistic models in an increasingly metaphoric way: He engendered tremendous interest outside anthropological circles, but alarmed more puritanical colleagues. (An anonymous reviewer in the London *Times Literary Supplement* on May 2, 1968, exhorted him to follow his anthropological predecessor Marcel Mauss, and not media analyst Marshall McLuhan.) Of particular concern during this period were the content of the "mind of the savage," his classificatory systems and his myths; in the wake of these studies Lévi-Strauss reached conclusions about the form that all human thought must assume.

Lévi-Strauss has forcefully argued that the "raw mind" operates in essentially the same manner, whether it be found in a Gallic savant or a Gilyak "savage" (the word favored by Lévi-Strauss). These individuals appear as mere receptacles into which minds have been placed. Lévi-Strauss compounds this somewhat mystical view with his assertion, unsurpassed in its combination of fatalism and hubris, that "in the end, it does not make much difference whether the thought of the Latin American native finds its forms in the operations of my own thought, or if mine finds it in the operation of theirs. What does matter is that the Human Mind, unconcerned with the identity of its occasional bearers, manifests in that operation a structure which becomes more and more intelligible to the degree that the doubly reflexive movements of two thoughts, working on one another, make progress" (McMahon, 1966, p. 58).

Echoing René Descartes, Lévi-Strauss holds that "I see clearly that there is nothing easier for me to understand than my own mind" (Descartes, 1960, p. 33). Though Lévi-Strauss's later asides have raised more than a few collegial eyebrows, he seems only to be making explicit the assumption under which any theoretical scientist works: that the human mind can structure and order data, whatever its source. Indeed, such a claim echoes Lévi-Strauss's conclusion in *The Savage Mind*: All minds, primitive or modern, order their universes in formally similar ways. So-called "primitives" may attend more to sensual properties, smells, shapes, or the distinctive "feelings" of things, while the Western scientist is likely to find his classifications upon latent or structural characteristics. But both forms of classification aim to be consistent, exhaustive, and acceptable to a community.

Among nonliterate cultures, the realm most closely related to science is myth. Lévi-Strauss's magnum opus is an attempt to demonstrate the elaborate, systematic nature of an entire myth corpus. Within a broad geographical region, a limited array of mythical characters emerges. Among the Indians of Central Brazil, for example, jaguars, human progenitors, otters, tobacco, and penis sheaths crop up in myth after myth. The relationships among such human, animal, natural, and cultural elements vary across myths so that, as a whole, the corpus explores all possible permutations of the dominant elements.

This variation is not performed solely for the sake of exhaustiveness, however; rather, the corpus addresses itself to a range of perennial problems, such as the qualities of family life, sexual ties, and economic activities, as well as the relationships between animal and human life, nature and culture, and life and death. The leading characters change, depending on the region and the preoccupation of a group; coyotes or tortoises may substitute for jaguars; axes or arrows for penis sheaths.

But two more important aspects of myths will persist: First, certain elements that cut across traditional classifications will serve as mediators, e.g., animals that can fly and live in the water (geese), and rites that transcend the bifurcation between nature and culture (the incest taboo); and second, certain conflicts or questions will be manifest in any corpus, because such concerns are universal.

Comment to readers in, 2024 or thereafter: I hope that you are still with me!

As myths serve no apparent function, and as they cover such an extraordinary sweep of topics, they provide for Lévi-Strauss what dreams provided for Freud and children's errors for Piaget: a novel, holistic, and occasionally poetic view of the human mind. Lévi-Strauss claims to be interested principally in the form of human thought; he has even introduced an algebraic equation, which pointedly disregards the specific content of myths. Some deem this formula meaningless, while many others, without prejudice to the formalism, find Lévi-Strauss's specification of the content of the mind the most exciting aspect of this work.

The work of Lévi-Strauss, then, is a daring (or, in the view of some, a foolhardy) attempt to specify both the form and content of various cultural systems: kinship, social organization, and mythic corpus. Because the basic structure of language underlies all such artifacts, a linguistic model can be used to elucidate them. Lévi-Strauss believes there is only a small set of possible codes:

> The ensemble of a people's customs has always its particular style; they form into systems. I am convinced that the number of these systems is never unlimited, and the human societies, like individual human beings, . . . never create absolutely; all they can do is to choose certain combinations from a repertory of ideas which it should be possible to reconstitute . . . one could eventually establish a periodic chart of chemical elements analogous to that devised by Mendeleyev. In this all customs, whether real or merely possible, would be grouped by families, and all that would remain for us to do would be to recognize those which societies had in point of fact adopted. (Lévi-Strauss, 1963, p. 58)

## IV

Though depending on similar tools, Piaget and Lévi-Strauss have invoked structuralism for different sorts of inquiries. Piaget has focused on the developing structures, the multifarious forms appearing en route to final equilibrium, while Lévi-Strauss has elected to describe the different forms the structure can ultimately assume.

These dissimilar endeavors reflect their differing points of departure. Lévi-Strauss's conception of language as the source and content of human thought has dictated an emphasis on the completed systems of thought, which have developed over the ages, are shared by all members of a culture, and can be simply transmitted to the young. Impressed no doubt by the ease with which a youngster learns one, or even several languages, Lévi-Strauss regards the acquisition of the code as no problem, but the relationship between various codes as fraught with puzzles.

Piaget's concentration on action, on the other hand, entails a concern with lengthy development antedating the ultimate forms (and formal operations) of thought. Systems like language, received essentially intact from the culture, may be parroted by young children; but without an understanding that only follows upon elaborate experimentation. Piaget's stringent criteria for understanding are epitomized in his laboratory interviews: The word *because* becomes mastered only when the child can employ it unambiguously in explaining various causal relationships.

In marked contrast, Lévi-Strauss has no interest in the accidental miscomprehensions of an isolated individual; he arrives at conclusions about thought from an examination of the complete, idealized system, reconstituted by the trained observer from numerous examples.

Such disparate orientations lead Piaget and Lévi-Strauss to divergent conclusions about primitive and Western thought. Lévi-Strauss strongly denies a qualitative distinction between the two forms; Piaget, though more cautious, suggests repeatedly that the "primitive's" thought ceases to develop beyond the level of concrete operations (thought dealing directly with environmental events); in contrast, Genevan adolescents proceed to formal operations (reasoning about propositions).

Both Piaget and Lévi-Strauss agree that primitive thought deals with the concrete; but while Lévi-Strauss deems such thought scientific, Piaget considers it an example of prescientific reasoning. The issue is explicitly joined in their references to language: Lévi-Strauss considers language the most important human capacity, one permitting classification, mythmaking, and explanation. Piaget considers language a source of confusions and speaks disparagingly of the "prelogic inherent in the usage of language" (Piaget, 1967, p. 191).

At somewhat closer range, then, our two thinkers appear to have different concerns and procedures. Piaget's concept of mind results from a *long history* of actions by a single individual, which leads that individual to increasing understanding of the structure of the world and the practice of science; Lévi-Strauss's concept of mind is a *given for all human beings*, a series of capacities immanent in the use of language, a rich depository of cultural systems, containing diverse contents, yet operating by means of identical formal mechanisms. Can those who are still haunted by the ghost in the machine somehow relate these distinctive views to each other?

## V

Piaget inveighs against the prominence of language, whereas Lévi-Strauss finds it the source of all wisdom. But neither man is particularly concerned with the manner in which a speaker has expressed herself (her syntax, nuances, or descriptions). Lévi-Strauss examines as many variants of the myth as possible, while Piaget poses as many questions (in as many different forms) as time allows. Both men are not so much interested in the manifest contents of the message, *but rather the nature of the reasoning or operations occurring at a level below language.*

Piaget examines a child's protocol and picks out the significant underlying propositions (which he can then order in the logical parlance of p's and q's); the mental action reflected in the protocol is a series of operations performed on the propositions. The individual has reached formal operations when he can systematically and exhaustively explore the relations between propositions describing a phenomenon.

Lévi-Strauss's procedure is comparable. Largely disregarding the manifest message, he also ferrets out what he regards as the principal propositions and sorts them according to similarities or relationships. This process yields a series of formulae; we learn that (a:b::c:d) one myth is the inverse of another; or that the message on the geographical plane reinforces those on the cosmological or social levels. A consideration of the entire corpus (which may run into hundreds of myths) reveals how a bevy of unsolved paradoxes and ambiguous characters emerging from the initial or reference myth have been thoroughly explored within the relevant folklore.

While the myth corpus might be said to illustrate formal operations (as it contains explorations of all possibilities), one should not conclude that the "singer of mythic tales" is capable of these operations. After all, an individual at the level of concrete operations can easily repeat specimens of formal thought prominent in his society. Should such a person prove able to relate the contents of one myth to that of other "transforms," or to create new myths on old themes, one would become more convinced of the formal nature of his thought.

Lévi-Strauss and Piaget employ linguistic materials in roughly parallel ways; accordingly, one is able to compare scientific and mythic reasoning, children and primitives, in ways satisfactory to the two thinkers. Such a comparison reveals that certain differences turn out to be more definitional than substantive; Lévi-Strauss's view embraces much more of thought, chiefly because he incorporates Piaget's underlying operations in his formulations about language.

While both thinkers disclaim interest in the specific elements about which humans reason, each inevitably becomes concerned with content. Indeed, Lévi-Strauss's treatment of content is a great strength of his system. He has persuasively demonstrated that the superficial variety of mythic topics actually exemplifies a limited number of themes. Of considerable import, too, is his acknowledgment that certain contents assume particular value. Thus, only singular creatures can qualify as myth mediators, for only a small number engender the "feel" of merging separate sensory or cognitive domains. Similarly, the specific names bestowed by "savages" transmit a magical aura or value to the phenomena named or to the namer.

In this insistence on the priority of particular objects or names, Lévi-Strauss seems to bolster Piaget's contention that thought is pre-logical unless it can transcend specific content. Yet the entire set of Piagetian tasks also involves the manipulation of content familiar to and valued by the Western child. Indeed, Piaget's own conviction that less familiar content must be assimilated to preexisting structures indicates that a child would be unable to exhibit formal reasoning in a domain where the objects or examples were totally foreign.

Lest one assume that Piaget's laboratory tasks are free of content bias, one must acknowledge that Western science represents a particular approach to knowledge; the Western child might well encounter difficulty reasoning about situations involving magic or kinship. Indeed, Piaget's work on moral judgment of Swiss youngsters demonstrates that reasoning that involves the different values attached to objects and to persons can be described in formal terms. Both Piaget and Lévi-Strauss are more concerned with content than they might wish, and neither has been able to overlook questions of value.

Piaget's and Lévi-Strauss's subtle approaches to the issue of value point up a critical difference between their enterprises. In asking what makes knowledge possible, Piaget is thinking of *scientific knowledge*; he wants to understand the nature of investigations in biology, mathematics, and sociology. As a result, he is primarily concerned with the relationship between the thinking, inquiring subject and the physical, value-free object. In this focus on subject-object relations, subjects are considered mature thinkers to the extent that they can treat objects qua objects, without the subjective overtones characterizing the thought of children and (Piaget would add) "primitives."

Lévi-Strauss, for his part, shows little concern for the world of objects, a peculiarly Western phenomenon. On his account, members of the cultures that he has studied tend to direct their greatest interest toward *kinship relations*; even when they exchange objects, these are regarded as mere tokens of the giver or the receiver. Lévi-Strauss typically examines the relationships between individuals, while Piaget examines the action of one person on the world of objects.

If there is one bias permeating the writings of the two men, it is a conviction that affective factors—feelings and emotions—cannot adequately explain *anything*. They seem motivated by a Gallic or Cartesian determination to leave the mind untarnished by basic desires, irrationalities, or affects. This perspective lends a certain one-sidedness to their accounts of the mind, and suggests two possible ways of supplementing their approaches: through Freud and through art.

## VI

A good physicalist, eager to uncover the material substrata of mental activity, Freud referred to subject relations as "object relations" and strove to view them dispassionately. All the same, one of Freud's signal contributions to psychology was his portrayal of subject-subject relations—the powerful feelings individuals evoke in one another and the influence of such passions on behavior and thought. Indeed, Freud maintained that if human desires were always sated, thought and reasoning would never develop.

More than a difference in emphasis separates the psychoanalyst from the structuralist. Neither Piaget's analysis of thought nor Lévi-Strauss's account of personal relations admits the possibility that the feelings held by individuals toward one another may color their perceptions, behavior, decisions, and evaluations. No doubt, the theorists would find it difficult to formalize such feelings, to specify their variety, to maintain that there is something "mental" rather than merely physiological about the affective life.

If the human mind is to be regarded as a computer, such an orientation may be justified; but developmental studies provide compelling evidence that the ways in which people feel and react toward one another, either instinctually or as a result of learning, alter their conceptions of every sort of object; in practice, they can never be disregarded. The motives leading to the investigation of certain scientific problems, the procedures by which they are probed, the manner in which mythical elements are chosen or kinship relationships are regulated, are all matters that a careful study of *subject-subject* relations might elucidate. Only an integration of structural and psychoanalytic approaches could reveal the range of affect dominant at each developmental stage, the influence of passion and rules upon one another, and the distinctive characteristics of subject-subject and subject-object relationships.

Freud, Piaget, Lévi-Strauss—all three of these imposing figures have exhibited ambivalence toward investigation of the arts. Each has flirted with the appeal of aesthetic objects or the temperament of the artist; none has directed his full analytic powers to such problems. If, overcoming their ambivalences, their disparate points of view could be cast upon the same questions, a meeting of the three minds just might be possible.

Let's conceive of the arts as specialized forms of communication. In these forms, multifarious aspects of one or more subjects (creators or performers) are conveyed through aesthetic objects (paintings, performances) to other subjects (the audience). Information communicated about the creating subject can be varied—his interests, choices, preferences, rhythms, emotions, or personality; but for our purposes, such information can be classified in either of two domains, depending on the nature of the art form.

In areas such as painting, where the artist expresses himself directly on the canvas that will be regarded by the viewer, the artist communicates aspects of his own behavior, such as the actions he has committed at a given historical moment or the way that these actions have been realized. The actual physical actions comprise an important component of the painter's style. In contrast, in other art forms, such as music or literature, no information is directly communicated about the creator's actions at a given moment; rather, the kind of information one gains about the creator inheres chiefly in the choices she has made from a code already extant in her culture, one that she transforms in some way to compose her particular message.

This distinction may indicate the kind of illumination one might expect from the three viewpoints. Our understanding of art forms in which action or behavior is predominant may be enriched by a cognitive "operational" orientation like Piaget's, while comprehension of arts in which choices are made from a preexisting language-like code seems more likely to follow from Lévi-Strauss's perspective. Finally, questions involving the relationship of the individual to an artwork—particularly the role of subjective experience in artistic creation and contemplation—may better be approached from the psychoanalytic perspective.

This brief foray into aesthetics prompts a recommendation: *Various orientations toward subjects and objects are more likely to be synthesized if they can be brought to bear on the same domain of experience, in which subjects communicate through objects.* One might draw upon the three sages for complementary explanations of how aspects of a person's acts or personality become expressed in an artwork; how cultural codes (including art forms and styles) are transformed and applied by individual artists; how aesthetic appreciation is possible; and how one learns to "read" an art object so as to avoid misleading information and glean meaningful insights about the thoughts and the feelings of the artist.

## REFERENCES

Anonymous. (1968, May 2). *London Times Literary Supplement.*
Descartes, R. (1960). *Meditations on first philosophy.* Liberal Arts Press.
Gardner, H. (1973). *The Quest for Mind: Piaget, Lévi-Strauss, and the structuralist movement.* Knopf.
Lévi-Strauss, C. (1963). *Structural anthropology.* Basic Books.
Lévi-Strauss, C. (1966). *The savage mind.* University of Chicago Press.
McMahon, J. H. (1966). Yale French Studies No. 36-37 (Translation of Lévi-Strauss, C. (1964). Le Cru et le Cuit. Pion.
Piaget, J. (1967). *Biologie et connaissance.* Gallimard.

# From Mode to Symbol

## Thoughts on the Genesis of the Arts

*While more of an empirical investigator than a theorist of the social sciences, I definitely have a theoretical bend. Accordingly, once I had decided on my career path, I read widely in the psychology and philosophy of the arts and began to think about the source of human interests and productions in the arts—dating back to the first years of childhood.*

*My first conceptual publication focused on this issue. And while I was already focusing on cognition, in the Piagetian sense, I actually drew on the psychoanalytic ideas of my undergraduate thesis adviser, Erik Erikson—another thinker who had considerable influence on what I studied and how I thought about it.*

*As befits a doctoral student, the essay was replete with references to earlier publications, most of which I have long since forgotten. But it does capture well what I was thinking about as I completed my doctoral studies. In these excerpts I present the core ideas—which I also explored in one of my first publications about imitation by infants (see Essay 7).*

Among the numerous problems that have intrigued aestheticians, few have proved more refractory to solution than the search for a common core to the several art forms and the attempt to pinpoint the genesis of human aesthetic activity. When problems have been so often posed without resulting in significant progress toward resolution, one may conclude that they are not susceptible of an answer or that a more appropriate version should be formulated.

Yet new insights may also be obtained if a question can be examined in a novel way or related to another line of inquiry. In this manner, some considerations from developmental psychology have suggested to me a way in which the nature of artistic activity might be illuminated and some progress made toward the resolution of longstanding puzzles.

Those intent on a conception of "art" that can encompass mime, music, and mosaics have not totally despaired; after all, the same adjectives or metaphors frequently seem appropriate as descriptions of phenomenal experience in these disparate realms. A *painting* or a *poem* may be perceived and then described as open or closed, direct or oblique, balanced or askew, penetrating, superficial, consuming, seductive, lofty or rough.

Though these characterizations rarely seem sufficiently precise, a lexicon of qualities has become our most effective means for describing the impact of an artwork. Reflection suggests that when adopting such labels, one frequently must search for the metaphor or simile that most accurately captures the sensations experienced in contemplating the work. Thus, in contrast to ordinary discourse, where the label appears as readily as one's "sense" of the object, we appear first *to experience the art work in an unlabelled way* and only subsequently find it necessary and possible to search for a description that can communicate our experience to others.

These two clues to a common core of the arts—the relevance of the language of qualities for characterizing aesthetic experiences and the initial nonlabeled reaction to a specific aesthetic object—are both suggestive of the early experience of the child: phenomena are apprehended nonverbally and category boundaries have yet to become fixed. Perhaps in considering the child's early experience and his first efforts to understand and communicate to others, we can receive further insight into the nature and development of the aesthetic experience.

What is known about the child's earliest experiences and feelings? Until the child can herself describe them, one must rely on inferences from her behavior, ranging from her physiological reactions to her facial expressions. The necessity of inference from ambiguous indices has not dissuaded psychologists from positing any number of "basic feelings" and from attempting full phenomenological descriptions of the child's affective life.

Despite some compelling guesses, however, the search for a set of childlike emotions and experiences seems destined to remain primarily a reflection of the psychologist's imaginative powers. On the other hand, recent efforts to characterize some of the overall features of the child's affective life, to focus on the general properties of feelings rather than on the identity of specific feelings, offer greater promise of verification.

An intriguing attempt to bridge the tantalizing gap between the child's physiological functioning and her psychological processes is Erik Erikson's description of the organ zones and modes. Proceeding from the psychoanalytic framework, Erikson (1950) outlines the manner in which libidinal energies are focused in the initial months of life upon the mouth, only to be displaced according to a predetermined plan onto the anal region, and subsequently onto the genital area. This much is common Freudian conviction; Erikson then proceeds to describe the manner or mode of functioning characteristic of each zone, drawing crucial psychological implications from the particular patterns.

Commencing with a description of the oral zone, Erikson records its preeminent function of ingesting food, at first rather passively and then increasingly actively and aggressively. Considering the purely reflexive of little psychological interest, Erikson directs his attention to two psychological modalities: "*getting,*" which derives from passive ingestion; and the more active "*taking*" or "*holding on,*" which derives from the search for food. The physiological has taken on psychological significance when the child not only experiences the ingesting of food but begins to "get" or "take in" actual experience, visual stimulation, parental love, objects, attention, and approval, and perceives a relation between these formally analogous activities.

Similar patterns are discerned by Erikson at the times when other bodily zones are in the ascendancy. The predominant modes during the subsequent anal period are *"holding on"* and *"letting go,"* primordially in relationship to the feces, with increasing frequency in relationship to other objects, individuals, and phenomena. So, too, the modes characteristic of the urethral-genital region—*intrusion, inception, and inclusion*—come to characterize the child's general approach to experience, figuring in his fantasy as well as his reality-oriented thought.

While each zone has its characteristic mode, other modes may also be exhibited there; the mouth and the "oral child" may be involved literally or figuratively in retention, elimination, or intrusion. Furthermore, the whole body or person may well come to be permeated by a particular mode; Erikson describes the child with heightened anal retention who can also become tight-lipped, expressionless, and rigid—and the patient who becomes "all mouth and thumb, as if milk were running through the body" (1950). The several modes each involve the relationship between an individual and the field or objects outside his body. Some emphasize keeping portions of the environment within one's own body, while others involve the release of what is or was once part of the body into the external field.

The modes may be regarded, then, as *a child's most primitive means of communication with the outside*. Despite these intermodal affinities, however, it seems most probable that each mode has a distinctive affective component. Observation of young and mature individuals suggests that passive taking-in of food and experience is somewhat different from active biting, and that elimination of waste or words is not identical with interjection. While the modes are akin to each other as various ways of relating the body to the world, differentiation between the principal ones seems indicated.

Erikson's contribution: *Physical modes gradually become social and cultural modalities, the child's way of experiencing the world beyond his skin*. A preliminary application of Erikson's view to the questions raised initially suggests that a child not only experiences the mode within his own body and in relationship to the environment in a subjective manner; *the child also tends to perceive the same kinds of modal configurations in various stimuli that she beholds and to behave in a "modal" way toward external objects*. For the young infant, the outside world is co-terminous with herself: At this point it is pointless to speak of "inner" or "outer" stimulation. As the child gradually differentiates herself from the environment, however, the sensations and percepts that have previously characterized the completely embedded self can be brought to bear on objects and forces that are acquiring a separate existence. Even as the child experiences modal functioning in her own body, she becomes sensitive to such properties as taking in or holding on in various individuals, objects, and works that she contemplates. Furthermore, the child's own activities in various realms, including manipulation of objects and symbol use, also realize modal properties that can be appreciated by the child and by others.

Modes, then, figure in the child's phenomenal feelings, her perceptions, and her creative activity. In marshaling evidence for these assertions, we turn to general findings of developmental psychology. Powerful insights into the cognitive and affective functioning of the child have resulted from study of the *errors or generalizations* that she makes. Evidence about the existence and significance of modal/vectorial properties (hereafter *m/v*) may accordingly be derived from this source. Even in the first year of life, the child is strongly influenced by many features of her environment, and her efforts to perceive, interact with, or imitate those features produce revealing behavioral patterns.

As examples, developmental psychologists Heinz Werner and Bernard Kaplan (1963) have been struck by the generalizations and analogies that the infant makes. Examples of infant imitation should indicate the extent to which the child appears to abstract out the modal and vectorial properties of his environment: We look as well to Jean Piaget's (1945) observations of his own children.

1. At 9 months one of Jean Piaget's children watches him stick out his tongue and then raises her forefinger. Piaget comments: "It would seem that the child's reaction can only be explained by the analogy between the protruded tongue and the raised forefinger."
2. At other times Piaget's daughter opens and closes her hand slowly after he has opened and closed his eyes; sticks her tongue in and out after he has alternated bread between his lips; opens and closes her mouth after he has opened and closed his eyes.
3. Piaget's daughter, while trying to extract a chain from a matchbox, slowly opens her mouth as if it aids her in locating the needed action- the widening of the slit.
4. Piaget issues forth a given number of syllables and his child repeats roughly the same pattern.
5. A 4-month-old child "began a curious and amusing mimicry of conversation in which she so closely imitated the ordinary cadences that persons in an adjacent room would mistake it for actual conversation."
6. Bernard Kaplan reports that his son of 13 months rocked his head from side to side in response to movements of other parts of his body; he also "accepted" equally his father's movements of head or arms as an adequate response to his own soliciting head movements.
7. My wife and I conducted a study of imitation in a 6-week-old infant. We found that even at that age the child was sensitive to the vectorial properties of extension/withdrawal and openness/closedness (see Essay 7).

Such instances reveal that the young child is sensitive to the *m/v* properties of behavior and stimuli; these properties impress themselves on the preverbal child and becomes abstracted from configurations as a basis for imitation. As Werner and Kaplan suggest, children were "sensitive to the dynamic properties of opening and closing and expressed these properties through different parts of the body capable of carrying out these dynamic features: the sensory-motor patterns possess qualities which defy a mere analysis of the movements of specific body parts; they have such qualities as direction, force, balance, rhythm and enclosingness" (1963, p. 86). It seems likely that a mode is experienced very early in a general way throughout the body, and that its vectorial properties become increasingly refined and differentiated in the course of development.

The trend from physiological (bodily) modes to psychological modalities of thought seems plausible to other observers. Thus psychoanalyst Lawrence Kubie maintains that:

> Since infancy and childhood, cravings arise in body tensions. It is inevitable that the child's thought should begin with his body and that his first concepts must deal with the parts, the products, the needs and the feelings of the body . . . the indirect representation of the parts of the body which are connected with our emotional and vegetative functions occur exceedingly early in the formation of language in the growing child . . . furthermore, they stand as a fringe and background tonus behind all conscious adult thought and feeling. (1934)

Writing from a somewhat different perspective, Konrad Lorenz notes the organism's tendency to generalize and suggests that responses to specific kinds of stimulation that have been programmed in the organism are soon extended to numerous features of the environment that share similar figurative properties (Inhelder & Tanner, 1956).

And Peter Wolff (1966), summarizing the developmental trajectory suggested here, indicates that

> congenital potentialities for psychic structure, each specific to one organ, become organizing principles for experiences related to other organ systems by mode displacements and become categories of thought independent of physiological activity by mode estrangement . . . the formal properties of action patterns rather than physiological activities per se are therefore the referent of the mode . . . what characterizes the mode is its configurational or space-organizing properties, which no doubt are derived from bodily function but extend far beyond these to become styles of veridical thought and social interaction.

The aforementioned examples and theoretical underpinning suggest that the modes and vectors are a flexible set of "dimensions" or "categories" through which the child initially experiences his own body and the world beyond. Precisely because of their generality, they are not restricted to one form of child behavior. Rather, one finds in the child's physical activity, his interaction with others, his imitation, his perception of various kinds of objects, and even his own manipulations and creations, clear signs of the organizing role played by the modes and vectors. The kinds of generalization and analogies made by the young child suggest further that even when particular modes or vectors appear in diverse domains, they are experienced as related in a fundamental way.

Such claims are certainly subject to challenge. One might question the extent to which the apparent prominence of modes and vectors is merely a reflection of the kind of language observers *necessarily resort to* in discussing activity and experience. Or one might concede the aptness of this language, yet claim it was so general as to make it difficult to conceive of analyses that would not concentrate on *m/v* properties.

These objections appear plausible. Indeed, we do talk of openness, penetration, depth, and so on in many contexts and feel that such properties permeate our experience. What is striking, then, is the extent to which such a tenor of discussion is almost entirely absent from any "hard-nosed" discussions of behavior in psychology. *The difficulty posed by modes and vectors is that they are "higher-level" terms that defy a ready analysis into simple quantifiable or physically controllable dimensions.* Most investigators have therefore shied away from *m/v* treatments because they despair of measuring reliably openness or closedness, potency or embeddedness, in an imitation, percept, or playful behavior. As a result, one infers from the psychological literature that head-turning, bar-pressing, and other conditionable responses are the child's chief activities; similarly, shape, color, and other manipulable stimulus being dimensions the aspects of configurations are most prone to be noticed. Yet observation of children in less structured situations suggests that these quantifiable and operationally attractive dimensions only emerge long after sensitivity to general *m/v* properties has dominated their experience.

Despite the various behavioral indices that point to the centrality of modes and vectors, it would be desirable to obtain less ambiguous indications of the child's capacity to note and to make use of *m/v* aspects of his experiences. For this reason, the child's *early use of language* provides a needed clarification of the way in which he groups and reacts to experiences. We note Piaget's report that his 14-month-old daughter used the sound "*tch*" whenever she saw a train pass by her balcony; yet she also employed the same sound when she saw other moving vehicles and people. The child appears to have seized upon the property of motion across situations and to be using "*tch*" to refer to events sharing this vectorial aspect. At 15 months the word "bow-wow" was said to anything moving; at 16 months, "no more" was used to apply to any type of going away; at 18 months "papeu" was used for anything going away; at 1 year "tata" applied to any kind of action. Piaget describes this early use of language as much more to systems of possible actions than to objects.

According to the general principles of development, an organism encountering a new experience or domain will employ the various adaptive mechanisms previously evolved before it slowly constructs new mechanisms more specifically suited for the particular system.

Hence, as children begin to encounter and to use human language, musical conventions, art objects, mathematical forms, and other symbolic systems, one would expect them to treat elements in these systems in ways analogous to those in which they previously treated other individuals, objects, and experiences. In contemplating the sounds of language and music, the sights of painting and sculpture, as well as in producing objects or patterns in these domains, I propose that children initially bring to bear their alertness to *m/v* qualities.

Some suggestive evidence:

- The earliest melodies spontaneously produced by children, for example, are principally falling minor thirds, which are succeeded by passages that rise and fall once and later twice. The directionality and sequence of the song rather than particular tones or melodies seem to be dominant.
- The child tends to link singing with his other activities in a way that highlights the common *m/v* properties; while singing he will dance, scribble, or rock, and the general quality of his tone will vary depending on whether he is running about the room, rocking back and forth, or teasing a friend.
- Depiction takes place not only through the exploitation of bodily movements but may indeed occur through extensions of bodily activity, for example, in drawings . . . the earliest forms of drawings are primitive scribblings caused by the arm movements projected on a surface . . . the later drawings of children still have this close tie to bodily equipment, but now the tie is to movements that are depictive of objects . . . their drawings appear to be translations of bodily gestural depictions of objects rather than representations of solely visual properties of objects . . . [one] child before copying a narrow triangle presented to him stretched out his tongue and then made rapid forward movements with his stretched-out forefinger; following this, he drew angular strokes so sharply that he tore the paper with the pencil. It was as if the presented triangle were grasped in terms of bodily movement—penetration, etc., and then drawn in terms of bodily depiction.
- Nowhere is the child's sensitivity to general patterns more strongly evinced than in his propensity toward *synesthesia*, his capacity to link elements perceived in one modality with elements perceived elsewhere to see sounds or to hear colors. For example, a 5-year-old recently listened to me play a variety of melodies on the piano. After each one he spontaneously named a letter of the alphabet, which indeed seemed "appropriate" for the particular tonal configuration.

The *m/v* hypothesis helps to elucidate the phenomenon by suggesting that the child does not perceive the stimulus configuration as an example of a traditional category (a high note of music, a yellow spot of drawing), but rather as a cluster of qualitative properties that could (and may) be equally well expressed and perceived through several modalities (something bountiful, brilliant, bluish). Some sort of structural continuity between diverse media and modalities appears to reveal itself to the child in a far more spontaneous way than is the case for the average adult.

Worthy of note, too, is the considerable accumulated evidence indicating that the art forms are at first undifferentiated in the child's mind and that singing and dancing, drawing and rhyming, etc. often occur at the same time and with matching $m/v$ properties. What might later be interpreted as inappropriate application of vectors (e.g., injecting the temporal element into the visual arts) seems a natural occurrence for young children, who prove extremely imaginative at cutting across standard dimensions.[1]

I suggest that early in life, when modes and vectors dominate the way in which the child functions, aesthetic experience (and other experiences as well) are permeated by, if not restricted to, $m/v$ qualities. As the child becomes an adult, however, he becomes increasingly acquainted with and influenced by the standards and norms of his culture, even as he acquires an increasingly sophisticated taste and makes stricter demands of the subject matter. Overlying the essential $m/v$ aspect of the art, then, are those cultural and cognitive components that differentiate, though they do not totally divorce, the child's experience of art from the adult's.

Aesthetic development consists, accordingly, in the gradual learning of cultural standards, mastery of necessary sensorimotor skills, development of a distinctive style, and an increasing ability to handle and appreciate complex presentations. These developments are hardly simple—indeed, they may span a lifetime—but they are essentially built on and continue to encompass the $m/v$ experience of art that characterizes the young schoolchild. The major developmental shift involves the ability to realize qualities in various symbol systems: Our evidence suggests that this process begins in the second year of life and is already well launched—particularly as regards the formal properties of the symbol systems—by the time a child is 7. Furthermore, for reasons not yet well understood, the trends are equally apparent in the child's creative and perceptual activities.

Since we have suggested that modes and vectors figure in all of the child's phenomenal experience, and continue to play a significant role for the adult, one might well ask why the above discussion applies particularly to the arts rather than to all areas involving symbols. As the human being can be considered a unified, integrated organism, it is evident—or at least arguable—that experiences in sciences, business, or sports also draw on $m/v$ aspects. I would suggest, however, that it is chiefly in the arts that $m/v$ properties are critical, constituting a

---

1. One must still decide whether the arts as experienced by the adult continue to draw on related kinds of psychological mechanisms and whether, accordingly, the developmental approach to aesthetics is revealing. While I speculated about possible answers to this question in this 1971 paper, much of our subsequent research at Project Zero tackled this question directly (see Essays 9–11).

large segment of the message and the significance of a specific object. Indeed, the sciences appear directed to the elimination of the ambiguity and suggestiveness involved in the general character of the modes, while the arts seek—indeed strain—to retain their allusiveness and semantic richness. Though one can approach scientific materials by focusing on their *m/v* aspects, or aesthetic materials by overlooking them, the overriding orientation of these domains would be undermined by such approaches. Just as the individual intrigued by *m/v* aspects of a theory is less likely to discard it in the face of conflicting evidence, the individual who has lost his sensitivity to these generalized aspects becomes an impoverished aesthetic observer.

In proceeding from a child who experiences *m/v* aspects in all realms to the adult participant in the artistic process, the individual comes to appreciate the distinctive aspects of particular realms and art forms. He no longer collapses arts and sciences, music and paintings, as does the youthful syncretic observer. Yet the particular sensitivity generally regarded as aesthetic resists a complete divorce of realms and continues to perceive affinities of the *m/v* sort between sensations, experiences, percepts, and activities. It is the continued alertness to general qualities—the attempt to embody and discern them in symbolic objects—that characterizes the aesthetic.

Though this essay has ranged widely, I have attempted to hold the guiding inquiries in mind. The nature of art may not have been elucidated definitively, but hopefully the suggestiveness of a developmental perspective has been conveyed.

Through the concept of organic modes and their vectorial properties, I have proposed that the early experience of the child has a distinctive yet consistent flavor that also dominates his initial contact with aesthetic symbols. This *m/v* approach to art becomes complemented by a greater cognitive complexity and an increased familiarity with established practices, yet it never is completely overwhelmed and thus remains in my view the bedrock on which aesthetic experience is based and the common aspect of all the arts. As modal properties characterize even the experiences of the first year of life, it is futile to point to a particular moment as the first "genuine" aesthetic experience. Yet the inquiry has suggested that the first use of symbols is already permeated by the modal aspect of the arts, and moreover, that the products and perceptions of the young schoolchild possess in significant measure the formal aspects of art. Requisite skill, experience, and intellectual development remain to be acquired before the child is a mature artist or connoisseur, but the modal sensitivity that belongs to him as a birthright must not be lost if he or she is to participate fully in the artistic process.

## REFERENCES

Erikson, E. (1950). *Childhood and society*. Norton.

Inhelder, B., & Tanner, J. M. (1956). *Discussions on child development*. Vol. 1. Routledge & Kegan Paul.

Kubie, L. S. (1934). Body symbolization and the development of language. *The Psychoana-lytic Quarterly*, *3*(3), 430–444. https://doi.org/10.1080/21674086.1934.11925214
Piaget, J. (1945). *Play dreams and imitations*. Norton.
Werner, H., & Kaplan, B. (1963). *Symbol formation*. Wiley.
Wolff, P. (1966). The causes, controls, and organization of behavior in the neonate. *Psychological Issues, 5*(1), monograph #17.

*Though my first published articles were more conceptual and theoretical than they were empirical or experimental, I was and remain a determined empiricist— and a sometimes experimentalist. Not surprisingly, when I became a parent, a sensitivity to modes and vectors (Essay 6) was very much on my mind.*

*While parenting Kerith, our first child, Judy Gardner and I observed an unusual phenomenon. In the opening weeks of life, Kerith was more likely to imitate the mode (open/close, extend/withdraw) than the particular part of the body that was involved (hand/mouth, finger/tongue). We collected data to support this form of imitation and published those data in* Child Development, *the leading journal in developmental psychology.*

*In the years following this publication, many other researchers reported similar phenomena. For a review of the literature, please see A. Meltzoff and H. K. Moore (1995).* A theory of the role of imitation in the emergence of self. In *The Self in Infancy: Theory and Research (pp. 73–93). Elsevier.*

# Note on Selective Imitation by a 6-Week-Old Infant

*Judith Gardner and Howard Gardner*

To determine whether a 6-week-old infant is capable of imitation, regular observations were made of one child's reactions to four kinds of parental behavior. Although there was little evidence of direct imitation, the data suggested that the child might be sensitive to the *modal properties* of the behaviors (see Essay 6).

When an action of an individual precedes a similar activity by an infant, and no efforts have been made to train the response, at least four possible explanations can be offered:

(1) the child may simply have been aroused and so responded with one of his limited repertoire of behaviors;

(2) the child may match the general rhythm or shape, merely because his attempts to assimilate the spectacle necessarily involve certain behavior with similar properties;

(3) the child may note the general vectorial modal characteristics (shape, direction, rhythm) and match these properties, yet not at the specific zone in which they have occurred; or

(4) the child may note both the modal properties of the behavior and the particular zone in which it occurred, imitating the behavior directly.

A number of observers have noted some imitative behavior in the early weeks of life, and some have even claimed direct imitation in the first month. Yet such behaviors have more often been dismissed as pseudo-imitation; and, in the absence of sufficient experimental data, most psychologists have favored the first or second explanation.

While observing our month-old infant, we became struck by the apparent specificity of her reactions and sought to make systematic observations. To achieve reliability, each author kept separate notes. We believe that, as there are but a few times during the day when the infant may be sufficiently alert to participate in such a study, parents are the most suitable experimenters. These findings are offered, less with confidence as to their finality, than in the hope of stimulating others similarly occupied to conduct such tests.

The data suggest that the extreme interpretations—complete imitation or simple arousal—are less probable than an intermediate interpretation. The partial imitation could be a function either of assimilation to the most convenient organ or the ability to abstract modal properties of the behavior. The finding that the child does extend and withdraw may be regarded as evidence against the "assimilation" and in favor of the "modal hypothesis."

To put this in lay language: The infant seems to be able to abstract properties like "open/closed" or "insert/withdraw" as well as—or perhaps even in preference to—a direct imitation of a specific body part.

## REFERENCE

Gardner, J., & Gardner, H. (1970). Note on selective imitation by a six-week-old infant. *Child Development, 41,* 1209–1213.

# Children's Sensitivity to Painting Styles

Every graduate student in psychology needs to find an area to investigate and make an initial investigation—otherwise, he or she will never receive a doctoral degree. As an enthusiast of the arts, I decided to examine the development of the ability to perceive stylistic features in works of art. And despite the fact that I am color-blind, prosopagnosic, and lack depth perception, I decided to focus on style perception in the visual arts. In this initial study, I established that only adolescents were able to overlook the content depicted in paintings and to attend instead to the identifying features of an artist's style.

*Summary:* A match-to-sample task was devised in order to determine whether children are sensitive to the styles of various painters. No significant difference was found in the performances of 1st-, 3rd-, and 6th-graders, but 9th-graders scored significantly better than younger subjects.

*Data:* On items where subject matter provided no cues, no significant differences were found between the highest- and lowest-scoring grades; where subject matter could be misleading, the oldest subjects performed significantly better. More effective problem-solving strategies, increased familiarity with art objects, and the ability to overlook subject matter and focus on aspects of technique are suggested as possible reasons for the superior performance of adolescents.

# DEVELOPMENTAL PSYCHOLOGY

Once I had determined to work for a doctorate in developmental psychology—whatever my other interests—I was expected to launch a program of empirical—read experimental!—research.

Given my interest in artistic development, I decided to embark on a study of how children become sensitive to the defining features of various art forms. And the concept that first captured my attention was artistic style.

Not only is style important in any art form—or, at a minimum, in any Western art form—but style can be captured and appreciated across the range of art forms. Accordingly, my colleagues and I embarked on a systematic study of how children come to appreciate artistic style. In this summary article, I present the results and conclusions from this line of research.

# Style Sensitivity in Children

*Abstract.* Traditional views of the process of style discrimination are presented and recent findings on the development of style sensitivity are reviewed. Style sensitivity in a connoisseur is seen as a complex capacity involving sensitivity to persons, sensitivity to objects, the discernment of gestalten, and the application of rules for discriminating among art objects. Factors affecting the development of style sensitivity are discussed, and a tentative model of the process is proposed. Style sensitivity is viewed as a type of concept formation, and the present studies are related to the theoretical approaches of Piaget, Bruner, and the Gibsons.

\*    \*    \*

The selection of tasks that reflect the power of human cognition, yet are amenable to controlled study, has proved to be a challenge to developmental psychologists. Frequently, tasks of concept formation have been employed; the concepts tend to be artificial ones, which do not exist outside the laboratory, are readily learned by adults, and involve the conjunction of specific dimensions, such as color, shape, or form. In contrast, most concepts utilized in daily activity are imprecise and defy ready definition.

It therefore remains an open question whether standard concept formation tasks model correctly the conceptualization that occurs during childhood. Study of a complex concept—one that figures in extra-laboratory (real world!) situations yet can be operationalized—potentially offers new insights about development. In this paper, style perception—a variety of concept formation that may occupy a central role in human cognition—is defined; conclusions from several studies are reviewed; and a tentative model of this form of conceptualization is proposed.

In what follows, style sensitivity has been operationalized as *the ability to group together works produced by one artist.* Tasks attempt to model the process where an adult correctly identifies an unfamiliar drawing as being by Leonardo da Vinci, or two Japanese works as being by the same possibly unknown artist. Numerous other perceptual activities, ranging from identification of a person by his handwriting to anticipation of the design of next year's Fords, would appear to involve some of the processes central in style sensitivity. In such cases, the subject is detecting subtle differences within a realm germane to classification of members of that realm.

To be sure, each form of recognition has its own features; recognition of blue jays differs from recognition of Picassos, Brahms, or cumulus clouds. Nonetheless, the resemblances among these mundane identifications may well be greater than the resemblance between any one of them and the sort of concept formation typically studied in the laboratory. In addition, sensitivity to style is an ability that develops to a remarkable degree in the connoisseur (thus allowing for the investigation of a high-level skill), yet is possessed in some measure even by the untrained adult. For these reasons, findings on style sensitivity should elucidate other forms of conceptual activity.

Until recently, discussion of style sensitivity was found principally in the writings of art historians and artists. Skill at recognizing an artist has been viewed as the product of many hours of intensive observation, coupled with attention to certain diagnostic features. While the style detector may explicitly look for specific details, much of his skill depends on a sensitivity to similarities of which he may have little explicit awareness. In style detection by the connoisseur, any cue is relevant: age of the canvas, slant of the signature, avoidance of certain motifs, jottings in a diary. Yet these features have been deemed less central to style detection than dimensions intrinsic to a given work: expressiveness, use of color or tone, texture, rhythm, and composition. According to the traditional view, a gifted connoisseur can "feel" himself into the artist's place, where he experiences the artist's emotions, ideas, and creative process.

As style detection in the adult is not well understood, attempts to investigate the development of this ability might be seen as premature. Yet, as Piaget has shown, an adult capacity may be clarified through an account of the process whereby this capacity evolves. On this argument, the nature of style sensitivity (and perhaps other conceptual behaviors) may be elucidated through a determination of the steps necessary for style detection by the young child.

Investigations of style raise numerous problems. Even after style has been equated with the work of a single individual, one must still determine whether all works of an artist are in the same style, whether the term should be restricted to works of one period, and whether there may be stylistic differences within a given work. Conveying the task to subjects may also involve difficulties, for young subjects have little if any understanding of the terms "style" or "an artist's work."

Accordingly, indirect methods of measuring style sensitivity must be devised. Such methods involve the assumption that cues within an artist's works permit stylistic groupings even by a person ignorant of the alleged commonalities among the products of a single hand. Other issues arise: Are abilities to recognize the style of an artist, to mimic it, to produce one's own style, all indices of style sensitivity? Is sensitivity to style closer to discrimination learning (as tested with animals), to conservation of structural properties (as in standard Piagetian tasks), or must it be viewed separately from other investigatory traditions? Finally, does style sensitivity possess commonalities across different symbol systems (music, words, pictures)?

In an effort to resolve some of these questions, my colleagues and I began several years ago to investigate the development of style sensitivity. The results of our studies can be briefly summarized. Because most of the research involved visual arts, the discussion will center on this area, with musical and literary studies mentioned when appropriate.

When preadolescents are asked to place together paintings that exhibit a common style, or paintings by the same artist, these subjects do not completely comprehend the request. Instead, they tend to sort paintings by their overall similarity, and this judgment is typically based on the subject matter of the painting. Works containing the same subject matter are assumed to be by the same artist, while works featuring diverse subject matter are assumed to be by different artists, even when they share compositional or expressive elements. Adolescent subjects, in comparison, are able to decenter from subject matter; the most successful ones exhibit knowledge of artistic traditions and periods, sensitivity to expressive and textural aspects of paintings, and an ability to weigh factors against one another

Subjects on the verge of adolescence are facilitated in style sorting if their attention is drawn away from the subject matter of the work; this shift can be accomplished through a presentation of works upside-down, or through elimination of works with deceptively similar subject matter. Such manipulations are not effective with subjects in the primary grades. Since these subjects consistently focus on the object being represented, it is important to determine whether they are nonetheless sensitive to nonobject cues (such as texture, composition, use of color) that are important in style detection, or whether, because of their singular involvement with subject matter, they fail to perceive these cues.

A training procedure was instituted. The goal was to determine whether preadolescents who are systematically reinforced for sorting by style, in the presence of deceptive subject matter, can learn to sort on that basis. Seven weeks of training resulted in a high level of style sensitivity in nearly all 10-year-olds tested and in a majority of 7-year-olds as well. Subjects trained to sort by style also retained the ability to sort by subject matter when the instructions were changed. Administration of a series of Piaget's concrete operational tests revealed that ability to perform concrete operations was neither a necessary nor a sufficient condition for style detection (see Essays 1 and 5 on Piaget).

In addition to demonstrating that 7-year-olds can attend to cues that mediate stylistic judgment, the training studies pointed up strategies that facilitate style detection. Subjects succeeded in *decentering* from the subject matter by considering the pictures as having "family relations" among one another. If two persons depicted in two different paintings looked alike, they were related to one another; or if two paintings with different subject matter nonetheless had a "family resemblance," subjects felt it legitimate to group them together.

Another aid to style detection was the subject's *hypothesis* of how the canvas, or the subject matter represented upon it, would feel if one could touch it. Paintings were grouped together if they were both bumpy, stringy, smooth, rough, evenly spaced, and so on. The manner in which pigment is deposited on the canvas appears to contribute to this aspect of an artwork and has led to a hypothesis: *Texture*—the microstructural regularities in a work that cut across figure and ground—provides an important cue in stylistic judgments. If two works, irrespective of different subject matter, share a common microstructure, they may well be exemplifications of the same style.

The contention that texture is a central contributor to style detection can be tested. Some support comes from the fact that subjects trained to sort by style scored somewhat higher than control subjects on pictorial discrimination tasks that involve textural judgments.

Our studies leave little doubt that subjects 10–12 years of age are sensitive to subtle aspects of a work of art, while raising the possibility that younger subjects are also alert to these cues. Studies employing works of art from other media, such as literature and music, suggest that the preadolescents may be as sensitive, or even more sensitive, to stylistic aspects than older subjects. Asked to complete stories that began in a certain style, and rated on the style sensitivity, originality, memory for details, and several other criteria, 6th-grade subjects displayed more literary skill than a matched group of 9th-graders. And, in a test of sensitivity to musical style, 6th-graders received absolutely higher scores than 9th-graders and college sophomores.

A consideration of the strategies used by the 6th-graders and by older subjects in the music task suggested that the two groups were approaching the tasks in different ways. The preadolescents were approaching the works without an appropriate vocabulary and making stylistic judgments on whether two works "felt" the same, "seemed" the same, or affected them in the same way. Older subjects, on the other hand, tended to apply labels drawn from aesthetic analysis or art history to the works ("they're both Baroque," "that one is in sonata form," "the other was in duple meter") and to make judgments based on whether the same labels applied to two works.

While each strategy seems plausible, the tendency to verbalize about one's reactions to a work, and to make judgments based on these analytic constructs, may not be a wholly beneficent development. Categories imposed by the culture may blind the perceiver to aspects of the work that a less informed (but equally acute) subject will take into account. The results of these studies suggest the hypothesis that preadolescents may be especially sensitive to the characteristics of works of art, perhaps because they already understand the meaning of style, yet have not adopted standard ways of approaching and classifying works of art.

Now that some of the general features of style sensitivity have been reviewed, it is appropriate to examine more detailed aspects of this capacity.

In earlier studies of style sensitivity, attention to the dominant figure was confounded with the cue of subject matter. That is, when a subject grouped together two representations of mountains, this identification might have been based on the fact that both paintings featured mountains, or that both featured curved masses at the center of the canvas.

Accordingly, sets were compiled in which the cues of subject matter and geometric figure were pitted against one another. Subjects were asked merely to group those paintings they thought most alike. College undergraduates without special training in the arts divided their responses almost equally between subject matter and geometric form groupings, suggesting that a mature population was equivalently sensitive to both kinds of cues. However, elementary school children overwhelmingly sorted by subject matter. Only girls at the 12-year-old level and some high school students showed a consistent ability to group by geometric form.

This finding demonstrates that for most children, the identity of what is being represented, rather than its purely geometric features, is central. Recognition of a perceptual configuration as an object in the world appears to be a fundamental perceptual tendency upon which aesthetic perception must be built; in contrast, sensitivity to the purely formal properties of a picture is found only among more sophisticated subjects.

The strategies exhibited by subjects continue to be scrutinized, but enough is now known to suggest the steps involved in style detection. Such a list can serve as a tentative model, useful for devising further studies.

## STYLE DETECTION IN THE ADULT

While the studies are not directly concerned with the procedures of the connoisseur, they suggest that proficient style detection involves at least two assumptions: each artist has a repertoire of skills that he can put to various uses, and there is a basic or central core, the artist's distinctive "handwriting," that should remain detectable despite attempts to hide his identity or to parody another's style. This defining set of features, a reliable index to the identity of the artist, is difficult to describe verbally and may remain outside the connoisseur's explicit awareness, only coalescing after much familiarity with an artist's work.

In addition to intuitive familiarity with an artist's distinctive "gestalt," the connoisseur will also be aware in a more explicit way of particular features of the artist's style. This may include knowledge about the materials available to the artist, the period in which he worked, the subject matter he favored, the manner in which he signed his work, and so on. Such factual information can be particularly valuable when the connoisseur has had little experience with the work of a particular artist or when the work shares formal or thematic features of several artists.

The sophisticated style detector should be able to use these kinds of information in a flexible manner. Her sense of what an artist might portray, and how she would portray it, as well as her understanding of the creative process, can serve as a useful guide. The capacities develop over a lengthy period and presuppose a developmental history, which may take the following form.

Available evidence suggests that the young infant will regularly attend to certain sensory properties, such as loud noises, the contours and edges of visual stimuli, and certain strong odors. Within a few months of birth, the child becomes sensitive to configurations and can recognize a parent, a mobile, a familiar visual or auditory pattern. He is becoming sensitive to persons when he can distinguish his mother from other persons, sensitive to objects when he begins to respond appropriately to his bottle, his ball, or his room.

These two forms of discrimination become increasingly fine-grained in the years following infancy, with the child eventually distinguishing very similar toys or objects. He learns to recognize a large number of persons, and comes to associate certain objects, behaviors, and traits with them: a pipe with his father, a voice with his mother, a dress with his sister, a handwriting with his uncle. Recognition of one of these traces becomes tantamount to identification of the relevant individual. Further differentiation within these spheres depends only on continued exposure to objects and persons and an indication from the culture that these differences matter

Our observations suggest that by the age of 7, the child's abilities to combine these lead to *incipient sensitivity to style*. A 1st-grader will note that representations of the same object differ from one another and that representations of different objects stem from the same person. He will then conclude that one house was drawn by Billy and another by Johnny, because he has learned to distinguish on the basis of visual clues between Johnny-drawn and Billy-drawn houses. Because he knows that Johnny draws in one way and Billy in another, and can recognize the drawings of each, he has appreciated a central element of sensitivity to style. However, the child will not comprehend the word or the concept of *style* and will not yet fully appreciate that every artist or person has a characteristic way of producing art works. (In retrospect, in 2024 any statement about the source of a work of art would require a great deal of knowledge about how works can be produced or revised!—HG)

Also important at this age is the child's increasing sensitivity to subtler aspects of sensory displays, such as rhythms, modes, textures, expressiveness, composition, and so on. As with style, the child may lack explicit awareness of these dimensions, conceptualizing them chiefly in metaphoric terms; but increasingly he will bring these dimensions to bear in his discriminations. Systematic focusing may lead to more explicit awareness of these dimensions.

What the prospective connoisseur still requires is a more explicit awareness of the dimensions that enter into style; the ability to search in an orderly manner for relevant cues and to apply these in making a judgment; and an understanding that individuals have characteristic ways of producing works. Knowledge of the history of an art form and biographical information may aid style recognition; yet the proclivity toward categorization and explicit hypothesis-testing may sometimes impede this capacity. Flexibility in approaching the works, a lack of preconceived and rigidly drawn categories, may permit the subject to immerse himself more completely in the work, discerning dimensions overlooked by more analytically oriented subjects.

A development that might enhance style sensitivity is the advent during adolescence of formal operations: the capacity to take all cues into account, to vary one factor at a time, to reason in a hypothetico-deductive manner. Yet a fundamental difference between the realm investigated by and those involved in style detection must be considered.

Most studies of concept formation involve "notational" concepts, which consist of a discrete set of isolatable elements (Goodman, 1968). In learning the concept of large green triangles, or in explicating the oscillations of a pendulum, the subject can determine when every combination of factors has been tried, and a formal account of his search can be written.

In contrast, the plastic arts involve innovational or "dense" symbol systems, which, as they contain an infinite (or at least indefinite) number of undifferentiable dimensions, cannot be broken down into basic units. No notation, duplicate, or score of a painting is possible because every nuance contributes in some way to the work and no exhaustive list of nuances is possible; accordingly, the power conveyed by a formal operational approach is of less moment. Nonetheless, it is still productive for the subject to consider a range of dimensions as long as he does not allow the quantitative result to dictate his judgment. An ability to work systematically with verbal propositions (rules) and to evaluate and contrast dimensions may still contribute to style detection. The skilled individual can devise a program for detecting the differences between individual styles and can apply it in a comprehensive way.

With an "open concept" like style, the program should be constantly changing in conformity with new information. The mature style detector appears able to express her strategy in rules, to adopt new rules, revise old ones, and arrive at a decision in a systematic way; however, this procedure should be complemented by attention to preconscious intuitions about the "feel" of an individual style. The child's openness to her own reactions and impressions is combined with the discipline and regularity of the intelligent adult.

Finally, the connoisseur should accumulate knowledge about characteristics of artists' styles and about factors relevant to style detection in general. While certain features may be diagnostic of a single artist's style, other features presumably characterize most artists working with a medium. It seems reasonable to expect *transfer* from skill at recognizing a few familiar artists to skill at recognizing an artist's style in general. The style detector draws on his experience in detecting familiar styles when confronted with unfamiliar ones.

To summarize our developmental model, then:

- The first stage in style detection is reached when the child's sensitivities to persons and objects combine, leading to awareness that a person's way of behaving will leave recognizable imprints in his creative products. At this time the child will also exhibit some sensitivity to general aspects of artistic works like expressiveness and rhythm.

- In the following years the child becomes increasingly aware of individual styles, facts about artists, and dimensions like texture and expressiveness. These sources can be effectively drawn upon as the art connoisseur achieves formal operations and develops a program for style detection. This program is enhanced to the extent that the consideration of dimensions is thorough, while open itself to change, based on fresh information.
- The ability to utilize one's own subjective impression of the "feel" of a work is an important supplement to this program; the connoisseur may on occasion reject the weight of conscious evidence and go with a feeling that is unsupported by logic or the most readily grasped features. It is hypothesized that this intuition is the result of years of careful observation of works of art—a "gestalt percept"—which, working implicitly, ferrets out the underlying constancies in the realm of objects Thus, the combination of capacities that characterizes a connoisseur reflects those unconscious processes that had impressed earlier commentators. . . . as well as the more explicit programs and categorical decisions that have emerged from psychological studies.

## PROBLEMS

The studies reviewed point to a number of areas requiring further investigation:

1. *Selection of suitable stimuli for the investigation of "open concepts."* Certainly, choice of materials is vital; yet in the absence of some method for classifying styles, the investigator risks using materials that will bias her results. One solution is to sample widely from the full range of available works.
2. *Difficulty of working with nonnotational symbol systems (Goodman, 1968).* Psychologists are understandably most comfortable with materials whose dimensions can fully be described and whose features can be duplicated. Predilection for notational systems imposes severe constraints on experimentation: the scientist refrains from using nonnotational kinds of materials and from postulating models that depart from the criterion of notationality. Given the evidence that cognitive mechanisms operate, at least in part, in a nonnotational way, it seems clear that any comprehensive cognitive psychology research program must eventually investigate nonnotational systems and develop effective methods of analyzing data from studies of these systems.

3. *Relationship between the ability to perform a task and explicit knowledge of the components of the performance.* In their daily behavior and under certain experimental manipulations, young children exhibit sensitivity to style; yet this is difficult to demonstrate so long as the subject lacks an explicit awareness of the concept of style. In many concept formation tasks, explicit awareness of dimensions is a prerequisite for a correct solution; yet insistence upon this rigorous criterion would result in an underestimation of sensitivity in the young child. In contrast to simple discrimination tasks, where verbalization is superfluous, but also in contrast to scientific problem solving, where explicit awareness of the relevant factors is required, tasks of style sensitivity occupy a middle ground, perhaps more representative of most daily cognitive functions.

## RELATION TO COGNITION

Two approaches are dominant in cognitive psychology: (1) that of J. J. and E. J. Gibson (1966, 1969), in which the increasing discriminations by the perceptual system guide the person to the discovery of aspects of stimulation that were originally not perceived; and (2) that of Piaget (1970) and Bruner et al. (1966), where the subject arrives at rules (derived either from his language or his actions) and applies these in his knowing of the world. In one case, the environment contains the information to which sensory systems gradually become attuned; in the other, the stage of development of the individual and the structures he has evolved determine what he perceives and how he interprets it. Information is in the subject's head, not in the world, which he can only know at the level of his mental structures.

I propose that style sensitivity is partially explicable in terms of each of these positions. On the one hand, the perceiver becomes more skilled at style detection simply because she has looked for many hours at works of art and so has come to appreciate distinctions and features missed by the untrained eye. Style, or at least texture, is a property of the stimulus, which any normal perceiver will eventually detect. On this interpretation style detection conforms to the Gibson model.

Paintings differ on an infinitude of dimensions, many irrelevant or misleading for style discrimination. As a result, style detection depends not only on refined differentiation but also on knowledge that individuals have characteristic ways of producing works; that works differ on explicable dimensions; that these features may be weighed against one another; and that various historical and biographical facts may provide a clue to the artist's identity. In this aspect, which involves strategies or operations (susceptible to formalization), the style detector resembles the models of cognition proposed by Bruner et al. (1966) and Piaget

(1970). She is a constructor of rules who allows her explicit knowledge to guide her perceptual processes.

Style detection depends in part on lengthy exposure to many works of art until an intuitive gestalt or an accumulation of assimilatory schemes has resulted. Yet this capacity is equally enhanced by the explicit search for relevant dimensions, and applications of rules that are not dependent on perceptions of the work.

As an example, recognizing continuities across Picasso's work is only part of style detection; one must also be aware that the artist's style changes in various ways, that others have drawn in similar styles, that nonperceptible factors can aid in a discrimination. Intuition that a work is by Picasso may be overridden by knowledge that he never used a certain material or that the canvas was painted over 100 years ago; here the Gibsons' approach does not suffice. Yet knowledge that a certain painter signed his name in one manner may have to be suspended when a careful gloss of an unfamiliar work leads to the conviction that only the particular artist under consideration could have painted the work.

In sum, then, style detection depends upon long-term exposure to works; the capacity to attend to relevant features and to overlook misleading ones—an issue investigated by theorists of discrimination learning; and the development of consistent strategies and search procedures that have been described by cognitive psychologists.

The studies and model described here help to modify the traditional view of style perception, while preserving those aspects that conform to empirical findings. That style may be a prototypical complex concept gains support from two demonstrations: (1) a combination of current theories is needed to provide a plausible model of the process; and (2) a comprehensive account of style sensitivity draws on a variety of perceptual and cognitive capacities, ranging from the most primitive perceptual capacities of the child to the most advanced probabilistic algorithms of the adult.

## REFERENCES

Bruner, J. S., Olver, R., & Greenfield, P. (1966). *Studies in cognitive growth*. Wiley.
Gardner, H. (1972). Style sensitivity in children. *Human Development, 15*(6), 325–338.
Gibson, E. J. (1969). Principles of perceptual Learning and development. Appleton-Century-Crofts.
Gibson, J. J. (1966). *The senses considered as perceptual systems*. Houghton Mifflin.
Goodman, N. (1968). *Languages of art*. Bobbs-Merrill.
Mussen, P. (Ed.) (1970). *Carmichael's manual of child psychology*, Volume 1. Wiley.

# The Development of Metaphoric Understanding

*Ellen Winner, Anne K. Rosenstiel, and Howard Gardner*

*To complement the study of style sensitivity, which took place mostly with works of visual art, my colleagues and I began a parallel set of studies with metaphor, a linguistic artistic form. In this article, Ellen Winner, Anne Rosenstiel, and I lay out the broad parameters of the comprehension of metaphor by young people. I reprint the "abstract" here, but it is rather abstract! The parameters of the study are described in the opening paragraphs, and the findings are conveyed in the discussion section.*

*For more details of this strand of work, please consult Essay 11.*

Two tasks were used to assess children's capacities to interpret metaphoric statements. Subjects ranging in age from 6 to 14 years were required either to explain a metaphoric sentence or to select one of four possible paraphrases. There was a developmental trend toward appropriate apprehension of metaphor.

Several steps preceding mature comprehension were described: metonymic and primitive-metaphoric interpretations were frequent prior to the age of 10, and the youngest children sometimes interpreted metaphors as descriptions of magical situations. In general, cross-sensory metaphors proved easier to comprehend than psychological-physical metaphors.

The results suggest certain cognitive prerequisites of mature metaphoric comprehension and help to clarify the ontogenetic relations among the abilities to produce, comprehend, and explicate the rationale of metaphoric language.

Results clarified the relationship between two types of metaphor. Primarily metonymic and primitive-metaphoric responses were offered by 6-, 7-, and 8-year-olds. Although no age group favored magical interpretations, children between 6 and 8 years had more magical responses than did older children. Those who gave magical responses did not take the sentences to be literal descriptions of the "real" world, but rather as descriptions of a fairytale world. However, there was evidence that these subjects were not entirely happy with this solution. One subject, for instance, said, "People can't become rocks! That's impossible!"

The conflict posed by *rejecting the logical absurdity of a magical interpreta-tion* and at the same time *honoring only the literal meaning of individual words* was solved metonymically by many of the children, who altered the expressed relationship from one of identity to one of contiguity.

By 8 years of age, subjects chose genuine-metaphoric responses as often as primitive ones; and by 10 years of age, subjects strongly favored genuine-metaphoric interpretations. The observed shift from metonymic to metaphoric interpretations may reflect the same processes that underlie the established trend in children from syntagmatic to paradigmatic word association.

Although 10-year-olds demonstrated a basic understanding of metaphor, they were often either unable to explain their interpretations or they had recourse to metonymic or primitive-metaphoric justifications. Similarly, these children some-times offered a genuine-metaphoric interpretation, followed immediately by a primitive one, suggesting the ease with which metaphoric thinking gave way to less sophisticated reasoning. An interesting (although not statistically signifi-cant) observation at this age was that 10-year-olds offered the highest number of inappropriate-metaphoric interpretations. Thus, these children exhibited the operations of metaphor before they could use words with precision.

A higher level of metaphoric understanding emerged in early adolescence. Whereas the 10-year-olds saw only one similarity between the two terms of the sentence, 14-year-olds could characterize the metaphoric relation in a variety of ways. For instance, a 10-year-old said that "The taste was a sharp knife" meant simply, "It was spicy," but a 14-year-old responded, "The taste was a shocking flavor, hitting all of my senses at once."

A comparison of the two types of items revealed that cross-sensory metaphors posed less difficulty than psychological-physical ones. The relative difficulty of the psychological-physical metaphors for younger subjects may have been due to an unfamiliarity with psychological domains; alternatively, the psychological-physical leap may be a comparatively greater one because the elements of cross-sensory metaphors are still both within the physical domain.

Whereas the developmental trends are relatively clear, it is more difficult to establish the strategies that led to the adoption of the various responses. It's not clear whether the "pre-metaphoric" understandings were due to an inability to interpret words on multiple levels (a linguistic immaturity); to an in-ability to perceive similarity between disparate objects (a cognitive deficit); to an insensitivity to cues indicating that a sentence is intended metaphorically (a problem of pragmatics); or to some amalgam of these factors and is a question open to further investigation.

The strategies that enabled children to adopt a metaphoric approach are also in need of clarification. The inclusion in each sentence of a dual-function adjec-tive meant that part of the resemblance between the two terms was explicit. Thus, the adjective *hard* may have served as a clue in finding a similarity between the prison guard and the rock; alternatively, the children may have ignored the reference to rock and simply found a way for the guard to be *hard*. Although both of these achievements culminate in a metaphorical response, the former is undoubtedly the more sophisticated.

Taken together with results of prior research on metaphor, these findings suggest that spontaneous production occurs first, followed by comprehension and then by the ability to explain the rationale of a metaphor. The spontaneous metaphors produced by young children are most often visual comparisons prompted by stimuli in the environment. In contrast, the comprehension of another's metaphor typically demands not only that both terms be imagined, but also that properties other than perceptual ones be taken into account. Finally, the ability to explicate the workings of a metaphor involves a distance from the processes of both metaphoric production and comprehension, as well as that metalinguistic awareness that typically only arises in preadolescence.

## REFERENCE

Winner, E., Rosenstiel, A. K., & Gardner, H. (1976). The development of metaphoric understanding. *Developmental Psychology, 12*(4), 289–297. https://doi.org/10.1037/0012-1649.12.4.289

# First Intimations of Artistry

*Howard Gardner and Ellen Winner*

*Given our interest in the origins of artistry in language (metaphor, stories) and in the visual realm (paintings and drawings), it became appropriate and timely to tie these lines of work together as appropriate. In this summary, Ellen Winner and I reflect on the common features, as well as the distinctive features, of early language and early depiction. And here we develop for the first time a perspective on artistic development that is U-shaped: early artistry is widespread; in middle childhood, a period of literalness (antithetical to artistry) ensues; and then, for just a segment of youth, there is a rejection of literalism and a return to a more artistic frame of mind.*

*Once I drew like Raphael but it has taken me a whole lifetime to learn to draw like children.*

—Picasso

## THE ENIGMA OF EARLY ARTISTIC PRODUCTION

Once dismissed as mere child's play, the drawings and utterances of young children are now treated with utmost seriousness. In part, this new respect for what was once considered idle scribbling or babbling reflects scholarly enterprise; students of human development are interested in the early manifestations of all human activities and therefore can ill afford to ignore any behaviors frequently engaged in by the young child. However, the fascination with what children draw and what they say goes beyond mere academic detective work. Both developmental psychologists and interested laymen experience genuine aesthetic pleasure from the drawings and the linguistic expressions of children. Indeed, it is now common to notice resemblances linking the graphic and verbal output of the preschool child to works produced by acknowledged adult painters and poets.

Testimony from both psychologists and artists has pointed to an early period of artistic flowering. Observers have pointed to a time of linguistic genius in which the child exhibits a sensitivity to the connotations and sounds of words.

Similarly, in the visual arts, many observers have considered the first years of drawing to be a very special time, a period when the shapes, colors, and arrangements of forms on the paper exhibit a strikingly expressive quality.

Intriguingly, however, the child's early artistic flair may decline during the years of middle childhood. Unlike such domains as reasoning and problem-solving, where researchers have more often than not found a continuing enhancement of powers with age, the trajectory of development in artistic spheres proves far less direct. The period during the school years has been described as a "literal stage" of language use, a time when the outpouring of figures of speech associated with early childhood has virtually ceased. And the golden age of the visual arts also seems lamentably short-lived.

Children of school age begin to spurn the originality and freedom of their earlier drawings in favor of an insistence on a conventional representation of the world about them, displaying a preoccupation with literalism that soon pervades (and tends to mute) their artistic consciousness.

Within at least these two art forms, then, we have evidence of a curve of development that departs sharply from the usual pattern. Rather than steady improvement with age, one witnesses a decline in the middle years of childhood, resulting in both fewer works and a more conventional (and thus, we argue, a less appealing) mode of artistic expression. Moreover, for most children, this withdrawal from artistry is permanent, and creation of works in language or picturing ceases altogether. Except for a handful of future artists, artistic development has been aborted.

At first inspection, then, the area of artistic achievement is marked by a quantitative and qualitative dip in performance, one that is permanent for most individuals. Such unexpected dips in the area of human development call for careful analysis. We must determine whether the decline is genuine and, if so, which factors spawn this atypical developmental trend. If, however, the decline is only apparent, we must determine whether it in fact represents an occasion for reorganization and growth. Finally, it may well be that certain aspects of the decline are not indicative of a genuine decline, whereas other aspects do reflect an actual loss.

A developmental analysis of artistry must adjudicate among these competing claims. Because so little has been firmly established about the development of artistic capacities, the number of domains that can be surveyed in any depth is limited. But two facets of artistry plumbed in recent years appear to exemplify a dip in development. One such area, drawn from the realm of language, involves the production of figurative language—in particular, metaphor. The second area, taken from the graphic realm, involves the production of drawings and paintings. In what follows, we examine and compare the pattern of development encountered in each of these domains.

Finally, we attempt an assessment of the extent to which early artistry bears genuine (as opposed to superficial) affinities to those achievements of mature artists that are so highly valued within our society. Much of this later account will necessarily be speculative, but we hope at least to go beyond sheer speculation.

## EARLY METAPHORS

The spontaneous linguistic output of toddlers provides a field day for students of child metaphor. Simply keep a notebook or turn on a tape recorder in the presence of a 2- or 3-year-old, and you will soon amass a long list of nonliteral uses of language—ones that can be shown to be genuine metaphorical rather than overextended uses. Listening to our own subjects, we heard a folded potato chip called "a cowboy hat," a streak of skywriting "a scar on the sky," and a group of nuns in their black and white habits "penguins." Studying the transcripts of spontaneous speech collected by Brown (1973) in his study of syntactic development, among the numerous examples of metaphor we found were a yoyo held up to a chin and called "a beard"; a peanut shell being pried open, called "a crocodile's mouth"; and black curly hair called "dark woods." And even Piaget (1962), hardly attuned to the artistic side of development, provides numerous examples of possible metaphors in the protocols of his own children: At about 2 years of age, Jacqueline, filling her hand with shells, reported "flowers"; putting a shell on the end of her finger, she called it a "thimble"; and, holding a brush over her head, she announced: "It's an umbrella."

Spontaneous metaphors produced during episodes of play constitute some evidence in favor of the young child's artistic bent. We must determine, however, whether children can also produce such engaging figures under more controlled experimental conditions—only then can we rule out the possibility that these metaphors are accidental creations and probe the capacity (as opposed to the inclination) to produce metaphors. Accordingly, we have in recent years devised simple linguistic tasks in which correct performance requires the production of a figure of speech. In one instance, we asked children of different ages to complete a story whose ending called for the production of a simile and then judged the endings on their novelty and appropriateness. In a second study, we gave children the opportunity to provide "pretend names" for a variety of objects—both ones for which they had names and ones for which they did not. Children were scored on their capacity to devise "renamings" that did not literally apply but that were appropriate rather than anomalous. For example, calling a flashlight battery "a sleeping bag all rolled up" was considered an appropriate metaphoric renaming.

These inquiries lent considerable support to the view of the young child as a competent metaphorizer. Indeed, scored on a blind basis, the endings and renamings produced by the young children proved comparable in amount and quality to those produced by older children and adult subjects. The ability to produce language that joins two usually disparate domains in a way appropriate for a given context seems available even to the preschool child.

It is possible, of course, that what counts as metaphor in adult speech is simply an overextension in the child's language—that is, a broadening of reference beyond the customary usage. Distinguishing between metaphor and overextension is tricky, and in this endeavor, one must steer between the risk of underrepresenting the child's knowledge (by calling all nonliteral uses of language overextensions) and overrepresenting what the child can do (by calling all metaphors). However, through the application of a set of converging criteria, which distinguish between metaphor and overextension (criteria such as previous literal usage of the terms in question, playful attitude, presence of symbolic gesture, etc.), we were able to determine with some confidence that a large number of early, nonliteral uses of words are indeed metaphoric (cf. Winner, 1979, for a detailed discussion of the criteria used). Moreover, we also observed that the preschool subjects in our renaming test based their pretend names on the same grounds (primarily shapes) and exhibited much the same pleasure in producing metaphors as did adults. All indices cast doubt on the claim that the child is engaged in an activity that differs fundamentally from that of the adult maker of metaphors.

Nonetheless, before we can cast the young child in the role of a full-fledged metaphoric being, we must come to grips with some apparently discordant data. First, we have the observation from a number of investigators that the production of creative figures of speech in spontaneous language does not last indefinitely; indeed, at about the time of school, the number of figures of speech drops quite dramatically. Considerable anecdotal evidence also points to a change in attitude with respect to figures of speech. Whereas young children appear to enjoy the transgression of boundaries entailed in making metaphors, school-age children avoid metaphors in their own speech and often express discomfort when such figures are produced by other individuals. There may, then, be a genuine loss of metaphoric flair during the school years.

A second problem hampering the view of the young child as a competent metaphorizer comes from literature on the comprehension (rather than production) of metaphor. A number of studies have documented considerable difficulties on the part of children in the understanding and explication of metaphors produced by others. Asked to paraphrase a psychological metaphor such as "The prison guard was a hard rock that could not be moved," one finds evidence of magical interpretations ("The prison guard was turned into a rock"); metonymic explanations, in which the terms of the metaphor are associated rather than compared ("The prison guard worked in a rocky prison"); physical interpretations ("The prison guard had hard muscles"); and erroneous psychological accounts that are almost, but not quite, correct ("The prison guard was 'stupid' or 'fussy'").

Moreover, even when given a set of pictures from which to choose the correct interpretation of a metaphor, children during the early school years often choose a literal one (e.g., a "heavy heart" is matched with a picture of a man lugging a valentine as often as it is matched with a tearful person). Such findings indicate that the capacity to comprehend metaphor is by no means present in the preschool years and may in some cases not be fully formed until the time of adolescence.

As one surveys this area, apparent paradoxes abound. Production seems to precede comprehension. Preschoolers name inventively in the same manner as adults. Older children behave less artistically than younger children. Can any sense be made of these apparently divergent findings?

A first step involves careful examination of the data on which some of the above claims are founded. It is true that in finishing a simile, preschoolers produce as many good novel comparisons as do adults. In fact, they produce an absolutely higher number. However, these subjects also produce a much larger number of inappropriate figures of speech, anomalous comparisons whose grounds cannot be decoded by adult judges. Thus, we must qualify our view of early metaphor by acknowledging that preschoolers lack that crucial "blue-penciling" knack—the ability to edit out metaphors that are idiosyncratic and thus noncommunicative—fundamental to any successful use of metaphor.

A second task is to distinguish between the loss of an inclination to produce figurative language as contrasted with the loss of an ability to do so. It is well documented that the spontaneous incidence of figurative language declines during the school years and that children of this age often reject figures of speech produced by others. However, our studies have shown that school-age children have not lost the competence to use metaphor. Given a "set" to produce figures of speech, or told that such a practice is acceptable in certain contexts, they typically prove able to produce figures of speech to the same extent, and based on the same grounds, as both their older and younger counterparts. Their reluctance seems to be a reflection of social and motivational factors rather than of cognitive limitations.

With respect to the finding that production appears to precede comprehension: probably the most important step in unraveling this paradox is to pay scrupulous attention to the kinds of metaphors at issue. Examination of the metaphors produced spontaneously by young children reveal that these figures invariably take one of a small number of forms. Either they are based primarily on the action taken by a subject upon an object (a child pulling a sweater up and down and then down again calls the sweater a "garage door"; runs a leaf through her hair and terms it a "comb"); or they are based primarily on physical resemblances between the topic and the vehicle (a red and white stop sign is called a "candy cane"; freckles are "cornflakes"); or they reflect a combination of action and physical grounds (a yellow pencil, which the child pretends to eat, is called "corn"; a bucket in which the child sticks a foot is called a "boot"). Virtually never heard are metaphors that exploit abstract, nonperceptual connections between realms, such as our prison guard example discussed above.

But it is precisely nonperceptual metaphors that prove baffling to schoolchildren. Indeed, those investigators who have required children to paraphrase or otherwise exhibit their understanding of nonperceptual metaphors are the ones who report predictable and consistent difficulties in decoding. We encounter a situation where those researchers who view the child as a metaphorizer focus on early perceptual or enactive metaphors; in contrast, those skeptical of the child's metaphoric inclinations dwell on the difficulties posed by conceptually based metaphors.

Indeed, once one takes into account the *kind* of metaphor, some paradoxes begin to dissolve. Children are unable to comprehend conceptual metaphors, and they rarely produce them on their own. It seems fair to conclude that this form of metaphor is simply beyond the ken of the young child.

Turn one's attention to other types of metaphor—to those based on striking perceptual similarities between topic and vehicle—and a far more impressive view of children's metaphoric powers emerges. We have already seen that young children readily produce these metaphors and have satisfied ourselves that such productions are, in all likelihood, not simply overextensions. Further evidence for the plausibility of our interpretation comes from the fact that, given metaphors that are simple-enough metaphors, even preschoolers are able to demonstrate incipient understanding.

Moreover, the nonlinguistic ability to perceive metaphoric connections—including ones that are based on abstract rather than physical grounds—may be present at an extremely young age. In a metaphorical "matching" task, preschoolers performed above chance, linking, for example, visual brightness to auditory loudness or tactile properties (hard, soft) to facial expressions in the same fashion as do adults. And, in a surprising finding in our laboratory, infants as young as 4 months have been found able to perceive links between metaphorically matched visual and auditory stimuli (e.g., an ascending tone and an arrow pointing up; a descending tone and an arrow pointing down). The ability to link stimuli on the basis of metaphorical similarities is, of course, only a necessary but not a sufficient condition for competence with linguistic metaphor. However, the fact that the perception of such similarities is present very early suggests an incipient metaphoric competence long before the emergence of the ability to understand metaphorical uses of words.

Our attempt to dissolve certain paradoxes surrounding early metaphor has met with some success. What is emerging is a more comprehensive and comprehensible picture in which young children are able to perceive abstract (nonphysical) metaphorical links between nonlinguistic stimuli; and in which preschoolers are able to both understand and produce linguistic metaphors based on physical similarities. Only during the middle elementary school years, however, do children demonstrate competence with linguistic metaphors based on nonphysical, abstract grounds.

Moreover, the decline in metaphoric production, although genuine, seems to be a decline in performance but not in competence. Children during the school years are able, when so instructed, to exhibit the kinds of metaphoric behaviors that we have come to expect from their younger counterparts—it is just that, on their own, they are unlikely to do so.

But have we succeeded too well in our task? Might there not be something special about young metaphorizers that still distinguishes them from older children and that constitutes a special kinship with the adult artist? We will return to this possibility after we have considered the nature of development in another artistic domain—that of drawing.

## FIRST DRAWINGS

At first glance, the case for youthful artistry in the graphic realm seems even more compelling than that claimed in the literary domain. At least in cultures that have been studied, most children draw. If given the opportunity, nearly every child produces hundreds of appealing drawings during the years before school. Moreover, these drawings exhibit more than a trivial similarity to those produced by leading artists of our day. Just one afternoon with a collection produced in an average preschool will yield a handful of drawings that bear a noticeable resemblance to such twentieth-century artists as Willem de Kooning, Helen Frankenthaler, or Jackson Pollock. To be sure, it is not difficult for a connoisseur to determine which were drawn by children and which by adult artists; what is noteworthy is that it is not ridiculous to suggest some similarities in these efforts.

Even more so than in the domain of metaphor, a dramatic change in their drawings follows children's entry into school. Children draw less and often seem to gain less pleasure from their drawing. Furthermore, the nature of their drawing undergoes a marked change. Earlier they were quite willing to experiment with colors, shapes, and composition—even when these efforts resulted in patterns never seen in the real world and rarely encountered even in the drawings of others. Now children impose harsh conditions on their artistry. Drawings must be representational; the more they approximate the culture's conventions of representation, the more these depictions are to be prized. Because there is little toleration for deviation from convention, unlike the preschooler, children of this age produce no abstract works of their own. And, unlike the preschooler, these older children also display a disdain for the abstract works of adult masters. Surely, here lies evidence of a developmental dip as dramatic as one is likely to find anywhere in the realm of cognition.

Yet whether these later conventional drawings indicate a decline in artistic quality is a question that proves challenging. Within the linguistic realm, where one can isolate discrete notational elements, one can examine the past history of word use and the child's own knowledge of word meaning and thereby assess the likelihood that an unexpected use of a word is genuinely metaphorical. Moreover, it proves possible to devise relatively simple tests of language production, thereby gaining evidence about whether the apparent metaphoric behavior in the child genuinely resembles that associated with an adult.

In a nonnotational art form like drawing, it is less clear what the units of analysis might be, or whether they even exist. Nor do simple paper-and-pencil tasks, which constrain the child's artistic behavior in an informative manner, readily come to mind. Yet, as we illustrated in the case of metaphor, unless one can devise some criteria for judging production as "artistic" or "nonartistic," the claim of the child as graphic artist rings hollow.

One basis on which to undertake such an investigation follows from philosophical analysis by Nelson Goodman (1968). In a probing treatise on the nature of artistic symbols, Goodman proposes a number of criteria exhibited by those symbols that are functioning in an aesthetic fashion. He indicates, for example, that artistic works tend to be replete: that is, they tend to exploit every aspect of the lines contained within them (as opposed, say, to a diagram or chart, where only the values of the ordinate and abscissa matter). Moreover, they tend to express properties that they do not actually possess (we often speak of a painting as sad or happy or busy).

These moods, it should be stressed, are not reflections of the artist's affective state at the time of production or of the audience's reaction at the time of perception or even of the subject matter being depicted. Even an abstract work can be "gloomy" or "lively." Rather, viewed within a symbolic context, these properties (like color and shading) belong to that symbol whether or not they can be perceived by every individual who encounters the work.

Armed with these definitions, we examined children's drawings in order to determine whether they indeed exhibited "symptoms" of the aesthetic (Carothers & Gardner, 1979). One possibility was to look at the spontaneously produced drawings and then to judge whether they exploited the symptoms of "expressiveness" and "repleteness." But this method could never offer conclusive evidence—it is not possible to determine whether features that appear replete or expressive to the observer were placed there deliberately or whether they were placed there unintentionally and accordingly could not be repeated. Therefore, we found it necessary to approach this issue in a less direct, but hopefully less ambiguous, manner.

We designed incomplete drawings identical to one another in all but one respect. To investigate sensitivity to repleteness, we prepared tracings of the same drawings that differed only in their use of line; the values that varied were thickness of line, brightness of line, and shading. Children were told the drawings had been made by different artists and were asked to complete them. We reasoned that a child sensitive to features of repleteness would be inclined to complete a given drawing using the same quality of line found in that drawing (a production task). And, just in case children are sensitive to such differences in line quality but are unable to control the quality of line in their own drawings because of motoric limitations, we also presented the children afterward with a multiple-choice task. Subjects were asked to choose between two given completions, one appropriate and one inappropriate (a perception task).

An analogous procedure was followed in examining the child's sensitivity to expressiveness. Two comparable drawings were prepared—both depicting a street scene with a man walking in front of a store. The drawings differed in the mood expressed—*happiness* in the case of one drawing, *gloom* in the case of the other. Once again, the children were asked to complete the drawing, this time by adding a tree and some flowers. An appropriate completion for the "happy" version might include some foliage that was upright, full, and curved gracefully; an appropriate completion for the contrasting "gloomy" version might include foliage that was bare, droopy, or in some other way stunted. Following their completions of these matched drawings, children were once again offered a multiple choice.

The results of this study were extremely straightforward. Very few 1st-graders showed sensitivity to either of the aforementioned aesthetic symptoms; they tended to complete both drawings in identical fashion and failed to choose the appropriate completion more often than chance. By 4th grade, children could generally choose the completion appropriate in repleteness or expression, although performance varied with the particular instance of repleteness featured. Perceptual success was high at the 6th-grade level, and productive capacity was reasonably good, although certain aspects of repleteness (e.g., the line quality of the shading) still posed a challenge for many of the preadolescents.

In this single study, thus, we find a strong suggestion that young children are *not* in control of the symptoms of the aesthetic. Their works may well be expressive to an adult audience, and they may appear to exploit properties of repleteness and expressiveness; but it is much less clear that children are exploiting these features intentionally. Moreover, these findings reverberate with others. A number of indices suggest that children begin to view works of art in a different way during the school years. Only at this time do they begin to attend to the style in which various effects are rendered in artworks: They study how an artist makes a car, a person, an animal; they examine the use of colors and shading; they ask for instruction in spatial depiction, perspective, foreshortening.

It would be possible for children to exhibit an increasing interest in the ways that effects are achieved without at the same time becoming fascinated with representationally accurate renditions. One can, after all, study the rendering in an abstract or highly stylized painting (this is actually done by gifted artists in adolescence and afterward). Yet it may be no accident that these trends—concern with realism and manner of rendition—co-occur in our society: that a concern with how effects are achieved coincides with an interest in accurate depiction of the surrounding world. But it seems safe to say, and consistent with available cross-cultural evidence, that children at this age strive to draw in the manner approved by their peers and their surrounding society. And since the manner most valued in our society is faithful representational depiction, it is certainly not surprising that most children embrace it.

Our experimental results, and our discussion, harbor a risk. There is a danger in denying that young children are capable of genuine artistry—the same peril that we encountered when we voiced skepticism about the special status of early metaphoric language. It is, after all, possible that our experimental manipulations—and they are no more than that—may circumvent children's natural métier or woefully underestimate their sensitivity to aesthetic features. Perhaps they have their own forms of expressiveness and repleteness to which we adults are insensitive. Or perhaps they could exhibit the very symptoms that seem lacking if we simply devised an easier task. Indeed, we have subsequently noted that, when explicitly asked to do so, a child as young as 6 or 7 can draw a "happy" or "sad" tree and children as young as 3 or 4 can draw a pair of lines that are recognizably "happy" or "sad."

To be sure, some of the causes underlying achievements may conceivably be artifactual. For example, a child may simply behave differently when asked to draw a happy line and a sad line. Nonetheless, if the resulting products can be distinguished accurately from one another by a blind judge (and more often than not, they can), we need to acknowledge at least an incipient sensitivity to expressiveness at an age far earlier than our studies have documented.

We will return to consider the possibly special outlook of the child at the threshold of school. To set the groundwork for this concluding discussion, we will first review some of the parallels and divergences between the profiles of literary and graphic spheres sketched in the preceding pages; then we then will place the artistic activity of 5-7-year-olds within the context of their general symbolic development.

## PATTERNS OF DEVELOPMENT IN LITERARY AND GRAPHIC REALMS

Our search for developmental patterns in the arts has touched upon two disparate domains. As representatives, respectively, of relatively notational and relatively nonnotational art forms, metaphoric and graphic performances span a spectrum of human capacities. We find, nonetheless, some striking parallels. In each case, the early years of childhood contain intimations of artistic behaviors. The young child produces numerous figures of speech, a surprising percentage of which seem tinged with metaphor; and, by the same token, this child fashions many drawings and paintings, at least some of which exhibit expressive power.

During the early school years, an apparent decline can be observed in both domains. Children draw less, and what they draw appears to be more conventional and thus less striking and original. Also, figurative language is often responded to by active rejection of any violation of conventional, rule-governed linguistic practice. Even if this "literal stage" constitutes a necessary step en route to the mastery entailed in developed adult artistry, it also marks the end of apparent artistic productivity in the case of most individuals.

But even as children's artistic productions hold less interest during the years of schooling, their comprehension of the artistic process seems on the increase. Only during this period do children become able to appreciate linguistic metaphors based on abstract grounds; and, in the graphic arts, only then do we find evidence of children's alertness to aesthetic qualities of works. Children are now able to respond to the stylistic, expressive, and replete properties of drawings even as they can, apparently for the first time, exploit such aesthetic features in their own work.

Although our review suggests that the aesthetic status of early figures of speech may be more secure than the aesthetic status afforded children's drawings, this finding may, at least in part, be an artifact of the particular methods used. When tasks are reduced to their simplest and most elementary forms—as in metaphoric-matching exercises or in direct instructions to draw in a certain expressive fashion—we find incipient signs of aesthetic sensitivity even in the preschool years. Surely the ease with which this sensitivity is elicited increases over the years, but some basic artistic sensitivity may be present at a much earlier time than our (and others') studies would suggest. To this possibility we now turn.

## PORTRAYING THE SKILLS OF THE YOUNG CHILD: A FIRST DRAFT OF ARTISTRY

Preschool children have seldom been characterized in a positive fashion. According to most standard texts, they are distinguished chiefly in terms of deficiencies: They are preoperational, illogical, and intuitive rather than rational. No longer susceptible to testing by means of the simplest "animal procedures" (where early intelligence can be exposed) and not yet able to respond to highly verbal tasks, these children seem to fail consistently on the measures devised by psychologists and, hence, to emerge as incompetent cognitive beings.

But perhaps nowhere in psychology do we encounter a greater disjunction between the authorized account of preschool children and their actual daily peregrinations. Watch a 3- or 4-year-old in our culture, and you will encounter a remarkably competent and energetic individual. At this age the child already talks in a highly fluent, adultlike manner; sings with increasing skill and clear signs of pleasure; and finds her way around, making wants known and solving problems encountered in everyday experience. Already across the threshold of representational drawing, the preschooler will soon be producing works of art holding considerable interest for others.

This account of preschoolers' rapidly expanding profile of skills has recently been corroborated by many researchers. Indeed, from a time when the preschooler has seemed bereft of abilities, we are moving to a contrasting state of affairs: a period when it seems increasingly difficult to distinguish the preoperational from the operational child. (See, for example, the subsequently published research by Alison Gopnik [1988, 2016].)

Our own cross-sectional and longitudinal studies have suggested that the central element in the child's growth during this preschool period is mastery of the symbol systems of the culture. At 12–18 months, human children still resemble intelligent primates—they are skilled in forms of sensorimotor intelligence, increasingly adept in tool use, but not yet able to represent to themselves aspects of their experience through one or another of the culture's symbol systems. By the age of 4, 5, or 6, on the other hand, they have attained a "first-draft" mastery of the major symbol systems: number, gesturing, oral language, storytelling, musical expression, dance, two- and three-dimensional depiction, and the like (see Wolf, 1979). They can produce legible instances in these symbol systems, decode them, and even display an emerging awareness of how these systems are best used or not used. In short, they have become symbolic creatures.

Attaining an initial fluency in such symbol systems is not the only accomplishment of this period. Many children—perhaps most—achieve this status with a speed that is truly remarkable and have transcended a simple fluency by the age of schooling. The developmental picture that has been so emphasized in the study of language—the rapid acquisition of phonology and syntax that drives some to posit an innate language acquisition device—may well have its counterpart in other symbolic realms. The rapidity with which children go through the various stages of scribbling and form-building en route to the attainment of representational skill, the processes whereby the child masters the musical scale of the culture, the relationships among the chief tones of the scale, the names and forms of scores of tunes—each of these suggest a computational capacity that requires relatively little environmental stimulation and modeling before it can "take off."

Moreover, in observing the symbolic flowering of the preschool child, one is struck by its intensive exploratory nature. In language learning, as illustrated by the pervasive nighttime monologues (Weir, 1962), children produce hundreds of sentence frames, trying out every possible combination of structure, and meaning. In the realm of graphic activity, children draw for hours at a stretch, experimenting with nearly every aspect of line, form, color, shape, and texture, and eventually attain a genuine feeling for composition. And even in the musical realm, particularly during times of solitary play or when falling asleep, children rehearse fragments of tunes and produce increasingly large segments of music that feature a coherence and structure of their own.

To be sure, such this activity is carried out with a relative lack of self-awareness. Indeed, while singing, babbling, or drawing, children sometimes appear almost in a trance and might well be unable to recall afterward just what they had been doing. But this, perhaps, is to the point. One detects in these activities a fundamental process of exploring perhaps crucial to the eventual mastery of the rules and conventions of the symbol system, one (perhaps fortunately) not entirely accessible to conscious control.

But to indicate the often-automatized flavor of this early "playing" activity is to relate only one side of the story. Equally striking, and equally relevant to the problem at hand, are far more systematic and self-conscious symbolic efforts in which the child becomes engaged. If children sometimes sketch or babble in a preconscious manner, at many other times they strive intentionally to express a certain idea, to reproduce a song that they like, to find a figure of speech that will capture what they want to say, to draw or model a specific person or object. If children sometimes produce one drawing after another in a seemingly haphazard fashion, they not infrequently announce what they want to draw, work assiduously to achieve this goal, express frustration when they fail, and often accomplish what they want and hand it proudly to a parent or teacher. In the preschoolers' exploration of symbol systems, intention, consciousness, and awareness are as much in evidence as automatic productions and preconscious rehearsals.

Moreover, part of the charm of preschoolers' symbolic activities may lie in their striving, with only partial success, to achieve precision, balance, and symmetry. Trying to emulate older individuals, they do not succeed, and their drawings are not thus perfectly symmetrical. But their valiant "preliteral" attempts share an affinity with the more deliberate asymmetrical creations of "postliteral" artists.

But how does this explosion of early creative activity—one that is being increasingly well documented—relate to the less dynamic aspects of artistry that follow during the literal stage? And how does it relate to the ultimate achievements of accomplished adult artists?

In our view, an understanding and appreciation of the nature of this early artistic activity—both what it is and what it is not—is crucial for any assessment of subsequent development in the artistic sphere. What seems at work during the preschool years is the attainment of an initial measure of mastery in various symbolic domains. Since children begin with no specific knowledge and yet can construct considerable knowledge and competence in a short period of time, their nervous systems must be designed to allow rapid and effective exploration of several such domains. This exploration—both conscious as well as unconscious—seems to give life to early artistic activity and may in fact yield the striking metaphors and drawings that have delighted so many observers.

Once this period of exploration has run its course, children have now attained a "first-draft" knowledge of artistry in several symbol systems. For a relatively brief period, they flower—at this time drawings and stories have their greatest appeal—and one can speak of a phase of genuine early artistry. But as quickly and brilliantly as this period erupts, it often seems to disappear again.

For, with the entering of school or its equivalent in "unschooled" cultures, children confront quite a different agenda: the need to master just what the rules of the society are, to follow them with fidelity, to be certain that one does not deviate, err, fail to conform. The very adventurousness that spawned such appealing works a few years ago is abruptly spurned in favor of a careful, almost compulsive, attention to detail; adherence to rules; and honoring of all the conventional boundaries. Now that children are intent on becoming full-fledged members of the society in which they are fated to live, it ill behooves them to tolerate that daring experimentation, that willingness to fail, of just a few seasons before.

For a small, though valued, proportion of the population, an alternative set of standards eventually come to the fore. Those who are chosen—or who choose themselves—to become artists begin to question the conformity of the elementary school years. They challenge the need to honor boundaries, the values that happen to be held by those about them, and the conventions governing the use of symbol systems. In their postconventional violation of boundaries, they recall preconventional young children, who, to the seeming disregard of the surrounding culture, follow their own muse.

But does the comparison between the young child and the adult artist withstand scrutiny? Certainly, it is not difficult to designate numerous areas of discrepancy between these two populations. Adult artists have passed through a literal stage: They have a conscious, articulate awareness of the conventions of literal word use and representational drawing and how and under what circumstances these conventions may be overridden. In contrast, preschoolers house a much less well-entrenched set of conventions; as a consequence, they presumably experience much less tension when they violate them or when they strive to achieve them but do not entirely succeed. When adult artists violate the standards held by other individuals, they do so by choice. Rarely are their deviations accidental.

Other differences also spring to mind. Adult artists are much more advanced technically, have far greater capacity to plan, will deliberately experiment in order to achieve certain effects, and are able to reflect critically on what they (and others) have done. Although not entirely absent, none of these virtues seems salient in the preschool artist.

And yet, the relations between the young child and the adult artist ought not to be minimized or trivialized. In both populations, one finds at work an insistent exploratory spirit; a willingness to ignore what others are doing and to pursue one's own personal agenda; a pleasure at the sheer (physical) act of creation and exploration; a willingness to try out diverse combinations, again and again if necessary, and not to rest until a desired outcome has been achieved. If this process is more self-conscious and deliberate in the artist (and it no doubt is), if the artist's goals are less fickle than the child's, the situation is by no means totally different for the younger counterpart: Even though child artists often give off clear signs that they are aware of the process(es) in which they are engaged, by the same token, adult artists oftentimes explore in a state best thought of as preconscious monitoring rather than focal awareness.

There is, we submit, yet a deeper link between the child and the adult artist. In both cases, one finds a profound reliance upon the art form as a means of self-expression. In the case of young children still uncertain about the concepts bandied around them, still inarticulate about many problems, fears, and wishes, the various symbol systems of the arts provide a unique channel whereby vague concepts and feelings can be brought to the fore for examination, articulated, and captured. The preschooler who relates stories about monsters, or who avidly sketches the protagonists of Batman or *Star Wars*, the child who draws over and over again a beloved animal—such children are exploring in the way that makes most sense to them the issues of greatest import in their lives.

The adult artist possesses the rules of literal language and, if pressed, may be able to articulate their concerns in such language. Yet—in the view of most students of artistry—literal words offer only an inadequate channel for expression of the deepest feelings and ideas. It is precisely because one is unable to "say it" properly in expository prose that the artist resorts time and time again to nonliteral symbol systems. It is through the use of metaphor, or through the arrangement of lines and color upon a page (or tones in a score), that the artist is best equipped to confront those thoughts that, while important, are not yet well understood.

Whether the rest of us are simply more successful in expressing equivalent thoughts exclusively in literal language, whether we resort to other vehicles (such as dreams) in lieu of an artistic gift, whether we have suppressed our desire to confront the artist's often threatening themes, or whether we simply have less to communicate in a compelling manner has scarcely been explored. Yet it is difficult to dispute the contention that young children and artists are drawn, in a way that most of us simply are not, to the exploration of nonliteral symbol systems.

## TWO FACETS OF ARTISTRY

Viewed from our vantage point, artistry is best thought of as consisting of two contrasting facets, each essential for any significant production. One has, on the one hand, the need for control, for standards, for rules. The procedures and standards whereby art works are produced must be well-known and understood, whether the artist intends to deviate only a little from these standards (as has traditionally been the goal of artistry) or to radically reject and transcend them (as has become the norm in recent decades). Accompanying this awareness of rules and conventions is the need for strong discipline. Only if artists are willing to come to know intimately the conventions of a symbolic system can they operate with facility within this system; and only if they know the rules can they transcend them effectively.

This Apollonian pole of artistry cannot suffice, however, if totally divorced from the spark—or demon—of creative energy. The need to explore, to revise, to tear down, to remake the world is equally part of the spirit of any artistic effort. If one adheres too closely to the rules, no achievement of significance is possible. One must have the desire and the will to try a new technique, to revise incessantly, to devise new means of expression when the present one fails or is found wanting. There must be a desire to encounter tension, to confront it, to overcome it, and to seek it time and again.

Here we encounter the special kinship of the child artist to the adult artist. For it is these two "species" of individuals who share the feelings of playful exploration, of total abandon, of dogged experimentation that constitute the other pole of artistry. It is possible, of course, that an adult artist might display these Dionysian tendencies without ever having encountered or experienced them as a child—possible, but extremely unlikely. It is more useful to view the feelings, experiences, and explorations of the child as a kind of initial investment, a first draft of artistry, one that remains in abeyance during the literal stage but that can (and will) subsequently be exploited by those individuals whose lot it is to become an artist. And, indeed, the challenge of art education may be to keep alive these Dionysian flames during the period when a more Apollonian approach to art seems unduly sanctified.

As we have noted, there has been a tendency in the developmental literature to *minimize* the accomplishment of the young child and to view all development as a continual enhancement of abilities. The appeal of this approach is clear: Researchers have come to expect, and to be able to measure, improvements with age.

As a reaction to this melioristic view of development, other researchers have claimed to find all abilities present at birth or unfolding with but minimal stimulation by the environment (Chomsky, 1975; Fodor, 1975). This search for knowledge in ovo is understandable, but to our view, equally untenable: The distinctions between the behaviors and capacities of the child and those of the adult are simply too profound to make this approach viable, except as a rhetorical ploy.

A third stream in the developmental literature is to plumb development for signs of U-shaped curves—areas marked by intriguing affinities between the early stages of achievement and those ultimately attained by masters. Some authorities emphasize the continuity between the two tips of the U, whereas others are more struck by the divergences.

Which side one takes in this dispute is, at least to some extent, a matter of epistemological scientific taste. By artfully changing definitions, by loosening or tightening criteria, one can render the child relatively more, or relatively less, like the adult. Similarly, by focusing on different spheres of development, one can also alter the coefficient of resemblance between the child and adult. Indeed, one major source of tension prevails between theorists (such as Piaget) who look primarily at scientific development and those (such as Erikson and Freud) who focus on the emotional life of the individual. These theorists may draw different conclusions about the relationship between childhood and adulthood just because, in the cognitive realm, the differences are more striking, whereas in the affective realm, parallels are far more telling, total reorganization far less likely.

Amidst these controversies, the arts provide an especially useful midpoint. As clearly as they involve cognitive skills, they draw with equal force upon the areas of affect, motivation, and will. And it is perhaps because they constitute a kind of *tertium quid* within the set of topics explored by developmental psychologists that they can serve as a fertile testing ground for competing theoretical claims.

We wish to note, in closing, the relativity entailed in any discussion of artistic development. *Our very conceptions of what constitutes art are constantly changing.* A comparison of child and adult art would hardly have been taken seriously 200 years ago, for the kinds of art prized by adults differed so strikingly from those produced by children that few individuals would cast a second glance at children's drawings. Possibly, 200 years from now, standards governing adult artistry and child art (and, of course, computer-generated art) will have come so close together that no one will even question the aesthetic status of children's works. Because standards may be more subject to change in the arts than in the sciences, the question of U-shapedness is more vexed in this than in other areas. It is a sobering thought that our view of what is developed, what is undeveloped, and which aspects of development are the "same" is not as inviolate and unchanging as we might wish them to be.

## REFERENCES

Brown, R. (1973). *A first language.* Harvard University Press.

Carothers, T., & Gardner, H. (1979). When children's drawings become art: The emergence of aesthetic production and perception. *Developmental Psychology, 15,* 570–580.

Chomsky, N. (1975). *Reflections on language.* Pantheon.

Chukovsky, K. (1968). *From two to five.* University of California Press.

Fodor, J. (1975). *The language of thought.* Crowell-Collier.

Goodman, N. (1968). *Languages of art.* Bobbs-Merrill.

Gopnik, A. (1988). *The philosophical baby.* Picador.

Gopnik, A. (2016). *The gardener and the carpenter: What the new science of child development tells us about the relationship between parents and children* (1st ed.). Farrar, Straus and Giroux.

Piaget, J. (1962). *Play, dreams, and imitation in childhood.* Norton.

Weir, R. (1962). *Language in the crib.* Mouton.

Winner, E. (1979). New names for old things: The emergence of metaphoric language. *Journal of Child Language, 6,* 469–491.

Wolf, D. (Ed.). (1979). Early symbolization. *New Directions for Child Development, 3.*

# Developmental Psychology After Piaget
## An Approach in Terms of Symbolization

*As is clear from my early writings, Jean Piaget had by far the greatest influence on my early thinking as a scholar. (In fact, I once quipped that if I had a choice of getting rid of Piaget, as compared to getting rid of the rest of developmental psychology, I might well choose to retain Piaget!) That said, as is true with most intellectual romances, one ultimately sees the weaknesses in the approach and begins to think about how one could continue the endeavor, with all due respect, while proceeding in new and often uncharted directions.*

*In this reflective essay (building on Essay 1) I delineate what should be retained from the Piagetian enterprise, what should be modified or tweaked, and what ought to be pursued in a new way. To be sure, I can't say that this recipe has been followed directly, even by my colleagues and me. In particular, the dependence upon stage theory, the notion that skills from different areas of cognition and involving different symbolic materials are closely linked, the neglect of emotional and (to some extent) social factors, and the exclusive focus on scientific (logical-mathematical) thinking have all been rethought. Whether I was simply reflecting an emerging zeitgeist, or contributing to the shift in some way, is not for me to judge.*

*Abstract*: While many of Piaget's contributions have already been assimilated into developmental psychology, certain limitations in his approach have become evident. Piaget's emphasis on logical-rational thought and his correlative neglect of the vehicles by which knowledge is carried point up the need for a new, post-Piagetian perspective. Here I outline an approach to cognitive development that builds upon certain of Piaget's assumptions and methods but takes into account the specific characteristics of diverse symbol systems and media. Such an approach may account for a number of phenomena left unexplained by Piaget, integrate diverse strands of research, and suggest certain promising new lines of investigation.

### INTRODUCTION: THE PIAGETIAN ENTERPRISE

Whatever its ultimate scientific fate, Piaget's contribution has over the past few decades provided a major impetus for research in developmental psychology.

Before Piaget began research and publication in the early 1920s, there was relatively little interest in the child's special cognitive and conceptual powers; most work consisted either of sheer descriptions of apparently objective features of the child's existence (physical milestones, preferred activities, motoric capacities); or anecdotal accounts of individual children, including ones displaying unusual abilities or difficulties; or broadly speculative interpretations of the course of growth. Moreover, when researchers investigated the child's mental capacities, they generally proceeded on the assumption that the child was simply an immature (or less skilled) adult (or, alternatively, a somewhat brighter animal).

Even after Piaget had begun publication, there were few who embraced his genetic epistemology (his term for his scholarly enterprise). Instead of exploring the ontogenesis of such fundamental (Kantian) categories of knowledge as space, time, or causality, and the relations among these domains, researchers continued to examine the manifestations in children of the classic topics of psychological analysis: memory, attention, perception, and problem-solving.

Piaget's major contributions—his stage sequence, his clinical approach, the fascinating phenomena he has uncovered—are well-known and widely respected (see Essays1 and 5). His overall enterprise seems more likely to survive than any competing approach in developmental psychology.

But the time has come to take stock of his bequest to the field and to consider its eventual fate. In what follows I will specify certain methodological guideposts that are central to the Piagetian enterprise; indicate a number of substantive problems in the Piagetian approach that have become increasingly apparent even to sympathetic observers; and then outline a new perspective that has recently emerged among scholars who wish to carry forward the program launched by Piaget.

## CRITIQUES OF THE PIAGETIAN ENTERPRISE

As befits an important but novel paradigm, Piaget's approach has been criticized from a variety of perspectives. At the extremes, certain investigators wish to return, wholly or in large part, to the epistemological positions that Piaget has explicitly rejected. One cadre of researchers embraces the empiricist position: Placing its reliance on the principles of learning theory, this group argues that the demonstrations made by Piaget can be adequately accounted for—and trained or extinguished—in terms of traditional principles of learning.

A rival set of investigators asserts that Piaget has, if anything, *overplayed* the role of interaction with the world in the development of mental capacities. Embracing a determinedly nativist account, such observers prefer to attribute to the child a rich initial mental structure that in itself contains the concepts, and the requisite representational machinery needed for eventual achievement of the Piagetian end states.

It is possible that these critiques will prevail, and that the Piagetian enterprise will be wholly abandoned, but I find this outcome very unlikely. Put briefly, the Piagetian enterprise has proved too fruitful, the competing stances too problematic, to make its demise probable. On the other hand, a number of deficiencies in the Piaget position have been so often noted that their eventual rectification has become a virtual certitude in post-Piagetian psychology. Among these gaps are Piaget's indifference to individual differences; his reluctance to deal substantively with issues of learning and pedagogy; his inattention to the role of specific sensory systems; his minimization of cultural and social factors in the child's mental development; his relative neglect of affective and motivational factors; his difficulty in explaining novelties in development (e.g., transitions); and his failure to develop an adequate conception of language, and of language's role in thought.

To my mind, there are two defects in Piaget's account that are critical, precisely because they lie close to the mission that he has explicitly pursued. First of all, even though Piaget has claimed to study the development of the mind, he has embraced a surprisingly narrow end state for cognition. In Piaget's view, mature cognition is no less, and no more, than the domain of logical-rational thought; accordingly, his end state is the competent scientist. Piaget consequently paid little heed to adult forms of cognition removed from the logic of science; there is scant consideration of the thought processes used by artists, writers, musicians, and athletes, and equally little information about processes of intuition, creativity, or novel thinking (see Essays 2, 3, 5, and 9).

The second central deficiency derives from Piaget's disregard of the particular materials, media of presentation, or symbol systems in which a task is posed, and a response secured. As far as one can ascertain, Piaget seems to believe that areas on investigation can be approached with equal vigor and accuracy irrespective of the physical materials used (beakers of water, balls of clay, building blocks, or billiard balls), symbol systems employed (language, pictures, gestures, or numbers), media of transmission (human voice, picture book, three-dimensional models), or mode of response tapped (verbal, pointing, sensorimotor actions, or some combination thereof). At most, Piaget is willing to concede a certain *"décalage"* across materials—that is, certain materials prove easier for children to work with.

Here Piaget has concentrated exclusively on the logical-rational thought of the scientist. Future investigators are likely to probe the skills needed by radically different kinds of thinkers. And where Piaget has rather cavalierly ignored the differences among media, materials, modes of response, and symbol systems, researchers are likely to attempt to unravel the specific effects of each of these elements.

Whether significant new insights are likely to flow from these studies will depend in large measure on whether such research efforts merely "tinker" with Piaget's basic demonstrations or whether they are informed by a fresh and fundamentally valid point of view. Because in my view such a point of view can be discerned in recent studies of human symbolic functioning, it is appropriate to turn to this area of study.

## SYMBOLIZATION: A STARTING POINT

Among developmental psychologists there is virtually universal agreement about a major transition in early childhood; in the second, third, and fourth years of life, the child moves from dealing directly (and exclusively) with the physical world of objects (animate and inanimate) to gaining, rendering, and communicating meanings through a range of symbolic vehicles. The 1-year-old knows a rattle, a mother, the Sun, only through direct sensorimotor experience; a 5-year-old can draw a picture, tell a story, and perhaps even offer gestural, numerical, or musical accounts of these topics. Theorists have used a variety of terms to characterize these capacities—symbolic, representational, semiotic, mediational, second-signal systems—but these authorities generally concur that this ability to refer to things and events at one (or more) symbolic steps removed is a hallmark of human cognition.

Offering a definition of symbolism is challenging. For present purposes, it may suffice to speak of a symbol as an element—usually a physical mark but one that (like a word) can also be an (abstract) conception: for individuals within a given culture, symbols carry meanings of one or another sort, and can generally enter into meaningful relationships with other elements from the same class of items, thereby constituting a symbol system.

The meaning relations are of two broad types: *denotational meaning*, in which the symbolic element designates an item that exists (or could potentially exist) in the world of experience; and *expressive* or *exemplificatory* meaning, in which the symbol captures a property (usually expressive, qualitative, or affective) of human experience without, however, designating a specific object or field of reference (cf. Gardner et al., 1974; Goodman, 1968, for elaboration of these points).

Where symbol systems differ instructively from one another is in the extent to which they embody the two principal forms of symbolic reference. Some symbol systems, like numerical notation, highlight *denotational* elements, while having little or no expressive potential; other systems, like improvised jazz, have only minimal denotational power but exhibit a wide range of *expressive* reference. Still other systems, like language, dance, or pictorial depiction, can vary from instances where denotation is key (e.g., scientific papers, mime, representational drawings and diagrams) to instances where expressive properties are foregrounded (poetry, modern dance, abstract art).

Rather than being essentially equivalent, symbol systems differ widely from one another in what they can encode (contrast language with music); what they typically express (contrast line drawings with dance); and which features they highlight (contrast the way in which volume is presented in number. music, or sculpture). Symbol systems also differ dramatically in their notationality: Some allow a faithful mapping back and forth between notations and a field of reference (e.g., musical scores, Morse code); others, infinitely replete, permit an indefinite number of readings, which thereby preclude any exhaustive mapping back and forth between elements and referents (consider a sketch or a dance performance).

Mastery of diverse symbol systems is a lengthy and complex process. One needs skills to read or decode symbol systems, to perform credibly within them, to relate the products of one symbol system (say, a diagram of a physical process) to those in another (a linguistic or numerical account of that process). Nor can we assume that information can ever be captured in exactly equivalent form across symbol systems: It may well be that each symbol system has its properties and limitations that preclude preservation of identity of meaning across an "intersymbolic" translation or paraphrase.

In light of these considerations, a program of research in developmental psychology suggests itself. The nature of competences in specific symbol systems needs to be defined; the steps en route to these achievements need to be delineated; the role of various symbolic competences in a wide range of adult end states needs to be specified; and perhaps most crucially, the relationships among these various trajectories of symbolic development and their connection to earlier accounts of cognitive development need to be articulated.

Clearly such a task will call for efforts of the range and perspective of those undertaken by Piaget in the logical-scientific domain. Of particular importance will be the delineation of the optimal units for the study of symbolism. It will be necessary to avoid the Scylla of equating all symbol systems (thereby returning in effect to an uncritically Piagetian position); but as well to avoid the Charybdis of spawning countless systems (and thus embracing a mindless "task" position wherein every use of symbols entails its own stage sequence). The aim, in other words, must be to slice Nature at its crucial "symbolic" joints.

Recognition of a range of symbols, and symbol systems, underscores the parochiality of Piaget's end state of cognition. For while numerical, and, to a lesser extent, linguistic symbols, dominate the everyday practice of the scientist, distinctly different symbol systems. and combinations of symbol systems, occupy center stage in other kinds of minds. A focus on musical cognition would highlight musical and numerical symbols; an interest in graphic cognition would stress pictorial symbol systems; an investigation of legal reasoning would highlight linguistic systems; and so on. Study of these diverse cognitive end states would likely yield different developmental trajectories and thereby call into serious question Piaget's univocal scheme of cognitive development: his belief—his faith—in "structures as a whole."

An approach in terms of symbol systems raises equally serious problems for Piaget' s assumption that all vehicles of meaning are essentially equivalent. Indeed, the basic thrust of the symbol system position outlined here suggests the following: Both the knowledge obtained, and the manner in which it is attained, *may differ fundamentally*, depending upon the specific materials, symbols, or media that serve as the vehicles of knowledge.

To be specific: Piaget would presumably contend that an individual's basic conception of time, space, or causality should be equally evident and comparably advanced, whether it be probed in words, pictures, or numbers. In contrast, the "symbol system" position holds that one's understanding of these spheres of knowledge will be fundamentally colored by the particular symbolic vehicles with which he has come into contact, and the particular materials and modalities with which this knowledge is tested.

As examples, one group of subjects might be expected to exhibit relatively advanced rhetorical causality (in the manner of a lawyer), while another group might excel in physical causality (in the manner of a scientist or engineer). By the same token, conceptions of time as embodied in music might turn out to be fundamentally different from the temporal concepts allowed in numerical or gestural systems.

Investigators of this ilk deviate from traditional cognitive-developmental theory in their willingness to treat as an empirical issue Piaget's assumption that a concept will be manifest in essentially equivalent form, independent of the specific symbol system employed.

In our laboratory we have approached this set of issues from two complementary perspectives. In a longitudinal study of a small group of preschool subjects, we have been examining the rate, and the manner, in which referential and expressive competences unfold in seven different symbolic media. A strictly Piagetian "semiotic approach" predicts that mastery of symbolic reference should unfold in closely parallel form across the range of systems; the rival "autonomy of symbol systems" perspective predicts little regularity across subjects and symbol systems.

We have in fact found evidence for an intermediate position: Close links exist among certain families of symbols (e.g., language is yoked to symbolic play, two-dimensional and three-dimensional depiction are closely allied); more distant links may obtain among certain other symbol systems (e.g., music and number), while little intrinsic relation exists among these three separate families (see Wolf & Gardner, 1979).

In a separate but parallel line of study, we have been investigating the breakdown after brain damage of the full range of symbol-using skills. These studies indicate that linguistic and pictorial symbolization occur in virtual independence of one another; that definite and unanticipated links exist among other more specific sets of symbol systems (e.g., ideographic scripts and pictorial representation; logical reasoning and comprehension of syntax); that certain other symbolic skills—like musical and numerical competence—seem organized in quite different ways across a range of subjects

Adoption of this perspective, it should be pointed out, often yields a focus on a set of issues different from those normally featured in Piagetian studies. A tension is created between the *concept at issue*—which can still be thought of as potentially representable in a highly general (logical) form—as contrasted with the numerous possible presentations. manifestations, and degrees of mastery of this concept. Attention comes to fall on the nature of specific materials and media—the skills needed to "read messages" within each of these media and to so employ materials that they embody messages that suit one's own purpose while also communicating effectively with others.

Indeed, the "disembodied" knowledge proved by Piagetians may actually be a fiction. Any concept, any domain, must be encountered in a specific medium, and in a particular task; understanding of competence inheres in a description of various "occasions for mastery" and a delineation of the kinds of transfer that can ultimately be expected. Indeed, it may well be that each material, or medium, has by its very nature a certain "symbolic" bias. There may be no way in which one can present absolutely equivalent stories in two media (say, television and book form): Each medium ineluctably favors certain symbol systems and highlights certain meanings at the expense of others. Such a state of affairs in no way lessens the interesting results that emerge when attempts are made to compare presentations across media; indeed, these findings merely reinforce the importance of attending not only to the symbol systems involved, but also to the particular media in which they are presented (Meringoff, 1980).

## RELATION TO OTHER LINES OF INQUIRY

Thus far, symbols, and symbol systems, have been portrayed as sets of meaning-bearing vehicles that can enter into relations with one another and can be mapped onto a field of reference. In practice, however, this description bypasses a host of crucial psychological distinctions. As suggested above, one needs to distinguish among symbol systems and (a) the media or technologies—the physical materials (like a television set or a book) in which a range of symbol systems can be conveyed; (b) the sensory and motor systems, which can likewise perceive and deliver instances from numerous symbol systems; and (c) the developmental domains, like map making or chess-playing, that may be peculiar to a specific culture and that, too, can invoke a number of symbol systems (Feldman, 1978).

A particularly daunting challenge concerns the relations between mental representations and the family of symbol systems. It may be that each symbol system has a specific, even unique mental presentation, but given the potential infinitude of symbol systems, it seems far more likely that families of symbol systems may be represented by related or even identical mental structures. It will therefore become important to build a model of the relations among these symbol systems: One will then have the opportunity to test predictions about which systems might complement or even enhance one another, which would be more likely to interfere with the operation (or development) of other systems. Such an inquiry may provide powerful evidence about the "natural joints" undergirding human symbolic capacities.

In conclusion: A principal task of developmental psychology after Piaget, then, is to identify and map the major domains of symbolic development; to trace the developmental milestones within each of these domains; to explore the relation that may obtain among these domains; to relate the emerging portrait of symbolic development to other traditions in developmental psychology; and, perhaps foremost, to ascertain the relation of the resulting portrait to the "central cognition" position outlined by Piaget. It is futile to anticipate in any detail the result of this enterprise, but it seems possible that there will emerge a number of domains with but a loose relation to one another, perhaps including a domain of logical rational thought (along the lines outlined by Piaget); a domain of visual representation; a sphere of language competence (combining syntactic, semantic, pragmatic, and figurative facets); and a separate avenue for musical growth.

Despite the argument put forth here, research on symbolic development may ultimately reinforce Piaget's overall picture, highlighting the principal operational structures and revealing only marginal modifications in terms of specific symbol systems. It may also emerge that we possess but a single language of mental representation. Perhaps, whatever the vagaries of individual development may be, all symbolic materials are eventually represented in a single computational lingua franca. Indeed, whether one chooses to highlight the degree of interconnectedness, or even identity, among the mental representations of particular symbol systems, or elects instead to highlight the importance of the discreteness of each symbol system, is at least to some extent a matter of definition and scientific strategy. Nonetheless, the time for a treatment in terms of symbolism seems at hand.

Ultimately, the role attributed to symbol use in cognition should not be entirely an accidental matter. It seems parsimonious to assume that we have been prepared by our evolutionary heritage to achieve competence in a range of symbol systems: In some systems, such as language, nearly everyone proves highly skilled. In others, like music, the distribution of gifts seems much less egalitarian. We cannot roll back the calendar of prehistory to discover whether the courses of individual symbol competences closely parallel one another, or whether each system has its own peculiar ontogenesis and genius.

Yet by examining patterns of breakdown of symbolic functioning after damage to the competent adult brain, one can receive intriguing cues regarding the range and the various organizations of symbol-using capacities. Such studies lend a measure of credence to the view that there exist overall levels of symbolic competence. But at present they offer much more sustenance to the rival position: Human beings have available a range of symbol-using faculties that draw upon a relatively independent set of cognitive skills. Indeed, the evidence emerging from neuropsychology may be the strongest grounds on which to base a belief that a study of the development of individual symbol-using capacities can significantly enhance our understanding of human cognition (see Essays 13–16).

## REFERENCES

Arnheim, R. (1969). *Visual thinking.* University of California Press.

Feldman, D. (1978). *Beyond universals. Mapping the terrain of development* [Unpublished manuscript]. Tufts University.

Gardner, H., Howard, V., & Perkins, D. (1974). Symbol systems: A philosophical, psychological, and educational investigation. In Olsen, D. (Ed.) *Media and symbols.* University of Chicago Press (pp. 27–54).

Goodman, N. (1968). *Languages of art.* Bobbs-Merrill.

Meringoff, L. K. (1980). Influence of the medium on children's story apprehension. *Journal of Educational Psychology, 72*(2), 240–249.

Wolf, D., & Gardner, H. (1979). Style and sequence in early symbolic play. In N. R. Smith & M. B. Franklin (Eds.), *Symbolic functioning in children* (pp. 117–138). Lawrence Erlbaum.

# INTRODUCTION TO THE STUDY OF BRAIN DAMAGE

During the early days of Project Zero, we invited scholars to speak to us about their work. At the time, we had heard about the exciting studies by Norman Geschwind, a professor of neurology at Harvard Medical School. He was one of the world's experts on the effects on human cognition from various forms of brain damage. It turned out that Geschwind's work cast light on an issue that Nelson Goodman and I had both been wrestling with—the extent to which artistic capacities are integral with, or quite discrete from, other cognitive capacities.

Geschwind's presentation was so riveting that he remained with us for several hours—and by the end of the evening, I had decided that, if possible, I would like to carry out postdoctoral research with Geschwind and to work with the brain-damaged patients under his care (see Essay 4).

A cruel fate of stroke, disease, or trauma, damage to the brain can provide vital insights into the organization of human cognition and behavior. An example: In normally functioning persons, it is extremely difficult to determine which capacities are identical or very similar to one another, and which are distinctly different in mode of operation and functioning. But through accidents of war, trauma, or aging, one can unravel human capacities and determine their relation—or lack of relation—to one another. As just one example, some cognitive scholars believe that music and language are quite similar perceptual and conceptual systems; others believe that they are quite discrete capacities. Studies of individuals with different kinds of cortical damage can reveal the features that are common to both systems, along with those that are specific to language, or specific to music.

For 20 years, I conducted neuropsychological research at the Boston Veterans Administration Medical Center. Day in and day out, I had the opportunity to observe the ways in which capacities break down—or are spared—through damage to the brain; and also, by implication, to discern which capacities are identical or very similar to one another and which ones are distinct. As was becoming my wont, in 1975 I recorded what I had learned in *The Shattered Mind: The Person After Brain Damage.* But I also had the opportunity to carry out various lines of research: observations of individual patients with unusual symptomatology, as well as more explicit experimental interventions with patients the nature of whose brain damage was known to us (at the time by brain and CT scan; now, of course, by more precise and more powerful imaging techniques).

In developmental psychology, I focused particularly on the development of artistic capacities; I followed a somewhat different route in neuropsychology.

On the one hand, in an effort to explain unusual symptomatology, I introduced certain concepts from Piagetian studies. Of particular interest was the notion of "operativity"—the extent to which one's ability to physically manipulate entities makes it easier to identify, name, and act upon those entities (see Essays 13 and 14). On the other hand, I looked at the breakdown under brain damage of certain human cognitive capacities—specifically metaphoric thinking (see Essays 15 and 16).

As it turned out, our studies illuminated previously undetected divisions of labor between the left and right hemispheres of the brain. Since the middle of the 19th century, it's been a mainstay of neurology that language functions are represented primarily in the left cortical hemisphere (in right-handed individuals—in left-handers, the situation turns out to be more complex). My colleagues and I were able to show that this claim is true for ordinary language functions (primarily syntax and semantics), but that when it comes to the pragmatic functions of language (nonliteral language, metaphor, irony, indirect speech acts, emotion), those capacities are significantly represented in the right (or so-called nonlinguistic) half of the brain. These insights, worked out initially with Ellen Winner and then with Hiram Brownell, constitute the empirical research in which I take the most pride.

For the 15 years following the receipt of my doctoral degree, I followed a wonderful though perilous course. I devoted all my working time to research! The wonderful part was being able to follow my curiosity wherever it took me; the perilous part was having to secure grants, to cover my own salary and that of any assistants, as well as supplies, travel, and other incidental expenses.

Fortunately, in those days it was far easier than it is today to get grants in the social and behavioral sciences.

So each weekday morning I traveled to the Boston Veterans Administration Medical Center in downtown Boston and worked with individuals who had suffered some kind of insult to their brain. Then, in the afternoon, I worked with young children, trying to understand their cognitive development (particularly in the arts). Sometimes this line of investigation occurred at our office at Harvard Project Zero in Cambridge; sometimes it occurred at local public schools in various Boston-area towns (again, it was easier to test in schools in those days than it is today—the idea of testing remotely, via computers, via Zoom, would have been seen as fantasy).

## REFERENCE

Gardner, H. (1975). *The shattered mind: The person after brain damage* (1st ed.). Knopf: distributed by Random House.

# The Contribution of Operativity to Naming Capacity

*As a dyed-in-the wool Piagetian at the time, it's not surprising that I donned Piagetian lenses when I began to work with aphasic patients. And in one of my first published studies with this population, I discovered that such patients had an easier time coming up with names of entities that could be easily handled and manipulated than the names of comparable entities that could only be known by sensory perception. Anyone who has experienced the tip-of-the-tongue phenomenon will likely recognize the utility of such tactile cues.*

## ABSTRACT

Matched groups of anterior aphasics, posterior aphasics, and a control group of nonaphasic patients were given an extensive naming test in order to determine the contribution of a number of variables to naming facility. Of particular interest was whether those elements described by Piaget as operative—objects, parts of objects, and other entities that can be readily grasped, manipulated, and operated upon—were easier to name than those that can only be known in a figurative way. When word frequency was controlled for, it was found that elements that are relatively operative were more easily named than elements that are relatively figurative. The results are discussed in terms of the mechanisms that govern naming in aphasic and normal individuals.

## DISCUSSION

It seems plausible to postulate that naming on confrontation involves the conveying of visual information to the cross-modal association area at the intersection of the temporal, parietal, and occipital lobes. At this point, all sensory and motor schemes aroused by the visual image of the object can be drawn on as the individual attempts to name it. In the case of letters, colors, and perhaps written words that arouse few if any nonvisual associations, the name is most likely to be evoked by a direct connection between the visual image and some kind of acoustic-motor engram. Naming of these written symbols (words, signs, notes) appears more susceptible to disruption by focal lesion or by cortical disconnections than naming of those visual images that arouse multiple motor and sensory association.

A similar argument may account for the relative robustness of the names of operative elements. These can be aroused through several sensory modalities, while the names of figurative elements depend primarily on associations within the visual modality. Further evidence of the contribution of operativity to naming are the findings that objects themselves are more likely to be remembered than pictures of objects or names of objects; and that children are more successful at naming operative than figurative elements. Perhaps this Piagetian concept will prove as fruitful in the realm of neuropsychology as it has been in developmental psychology.

# Bee but Not Be

## Oral Reading of Single Words in Aphasia and Alexia

*In a study related to the previous essay (Essay 13), Edgar Zurif and I studied the ability to read single words in three groups of patients: those with problems in spoken language (aphasics); those with selective problems in reading (alexics); and those with problems in both reading and naming (alexia with agraphia). When patients have preserved reading capacities, they can read equally well words with and without operative potential. But when oral language and/or reading are impaired, then the "operativity" of the entity being referred to contributes to its legibility. As the title signals, it's easier for such patients to read "bee" than to read "be."*

### ABSTRACT

Three studies are reported of the oral reading ability of language-impaired patients. Part of speech and picturability are shown to contribute to a word's readability. In addition, words whose referents can be easily manipulated (operative nouns) prove easier to read than matched words whose referents are relatively figurative.

BACKGROUND Comparisons of reading and naming ability were drawn between aphasic patients and patients clinically diagnosed as alexic. These comparisons suggest that for most aphasics, reading and naming are mediated by separate mechanisms, with the ability to read being relatively spared; that alexics without agraphia (i.e., loss of ability to read but not the ability to write) achieve most success with short words, irrespective of the part of speech; and that alexics with agraphia (loss of both reading and writing abilities) achieve most success with picturable nouns, even when these contain more letters or syllables. In alexia with agraphia, reading and naming occur at a comparable level and may be mediated by identical or similar cognitive mechanisms.

If oral reading in aphasia simply reflected what remained of the capacity to translate constituent graphemes into sound, then charting this ability would be of scant interest. However, the various ways in which aphasic reading breaks down seem to require a more complexly based level of analysis: Witness frequent reports attesting to the role of reference in oral reading, the patient: for example, not being able to read "to be or not to be," but succeeding with "2 bee oar knot 2 bee." It is quite likely, even in aphasia, that some form of processing or sampling for meaning is carried out, irrespective of the visual or acoustic properties of the specific words comprising the utterance.

We detail here some of the meaning-implicated variables involved in oral reading. These analyses of reading are presented both for aphasia and for those syndromes in which reading appears to be even more impaired than auditory language skills—the two classic forms of alexia. Also examined is the relationship between the language-impaired patient's ability to read aloud single words and his ability to name the visually presented referents of these words. Given the instructive manner in which these two processes normally differ and the fact that they can be differentially affected by brain damage, such a comparison might be expected to underline the critical variables involved—in effect, isolating those cognitive dimensions that are common to reading and naming, as well as those unique to each of the two processes. Moreover, the range of oral communication capacities may be elucidated by specifying the contributions of specific linguistic and cognitive factors to oral reading.

## FINDINGS

These two varieties of alexia bracket the mechanisms of oral reading. The alexic with agraphia has lost the ability to proceed from the visual grapheme to its corresponding individual sounds but retains the capacity to proceed from a familiar visual configuration to its meaning; the pure alexic has retained the capacity to match graphemes with sound but has lost the ability to view the word as a familiar whole, and to derive its meaning and name from its habitual association with a concrete object.

The pure alexic, then, has a difficulty essentially remote from aphasia: If the signal could be conveyed to him via another channel, his apparent reading difficulties would largely disappear. The alexic with agraphia, on the other hand, epitomizes the difficulty of the typical aphasic patient. He has lost the capacity to carry out the mechanics of reading while retaining the penumbra of meaning and the name surrounding more familiar and accessible words. Thus he benefits in particular from noun-ness (denoting a person, place, or thing) and also operativity (the manipulability of the object)—two factors that apparently elicit his remaining associations and lead him, if indirectly, to the desired oral target. His difficulty with the small grammatical words appears to stem from the fact that, though familiar, these words are likely to have relatively little meaning in isolation; they therefore fail to arouse the articulatory patterns involved in their utterances. Here, then, is a provisional explanation for why the word *bee* poses no problem for most aphasics, while the word *be* is very difficult for all except the patient with pure alexia.

## REFERENCE

Gardner, H., & Zurif, E. (1975). Bee but not be: Oral reading of single words in aphasia and alexia. *Neuropsychologia, 13*(2), 181–190.

# The Comprehension of Metaphor in Brain-Damaged Patients

*Ellen Winner and Howard Gardner*

*As an empirical investigator, in collaboration with others, I have carried out several dozen studies—some with children, some with patients who have various kinds of cortical injury, a good number with both populations. In my view, by far the most important and most original line of work involved the study on the breakdown of linguistic capacities in individuals with brain damage. While metaphor is clearly a linguistic function, Ellen Winner and I documented a deficit in metaphoric understanding in patients with right hemisphere damage, and a surprising preservation of sensitivity to metaphor thinking in patients with damage to the left hemisphere, which normally undermines linguistic capacities.*

*On the basis of this intriguing finding, we carried out other studies of language capacities in patients with right hemisphere damage, and many other researchers have also pursued this line of work. As is often the case, the role of the cortex in various kinds of linguistic capacities turns out to be more complex than we had anticipated. But science does not entail the final word—it entails opening new ground, which we did in this article.*

Left-hemisphere dominance for language (in right-handers) is the most well-established instance of division of labor between the cerebral hemispheres. However, this characterization rests almost entirely on studies probing the appreciation of language in its literal, denotative aspects. Whether the left-hemisphere dominance for language extends as well to other, more figurative uses has not been established.

The ability to transcend literal meanings proves essential in normal language use. For instance, in all languages, terms used to refer to psychological aspects of experience are drawn from the physical world (for example, a hard heart; a dry wit). In a similar fashion, percepts drawn from a particular sensory modality are described by terms drawn, synaesthetically, from another sensory domain (for example, a loud color; a bright smell). Where linguistic performance relies on only the primary denotative senses of words, communication is not only severely impoverished but, in many cases, actually prohibited.

While metaphoric expressions thus constitute a common part of our language, appreciation of such linguistic figures involves a number of discrete cognitive operations, such as the abilities to go beyond the literal and to perceive a similarity between alien domains. Several investigations have shown that children of early elementary school age have considerable difficulty in one or more of these operations (see Essays 10–11), difficulties that might well be exhibited by patients with organic brain disease. Accordingly, an examination of such metaphoric operations among brain-damaged patients may illuminate the manner in which this aspect of language is normally processed.

Even as the characterization of the left hemisphere as dominant for language has been widely accepted, analogous attempts have been made to characterize the principal operations of the right hemisphere. Frequently suggested are a minor hemisphere dominance for visual or visuospatial processing. Somewhat more speculative, but highly intriguing, have been descriptions in terms of creativity or aesthetic sensitivity.

An examination of the metaphoric competence of right-hemisphere-injured patients can help to adjudicate among these competing descriptions. On the one hand, because characterizations of the right hemisphere as dominant in aesthetic matters have relied almost exclusively upon results with pictorial materials, they have confounded the "visual" with the "aesthetic" hypotheses; accordingly, a task employing materials that are at once nonpictorial but aesthetic can more clearly evaluate the "aesthetic claim." At the same time, a study of metaphoric competence can shed light on other recent characterizations of the right hemisphere in terms of its centrality in emotional or situational sensitivity.

In the present study, two simple tasks of metaphoric competence were devised and administered to groups of normal and brain-damaged patients. In addition to clarifying the above characterizations of the hemispheres as "linguistic" and "aesthetic," the following questions were considered:

(1) What is the overall competence of normal adults and brain-damaged patients on a task in which a metaphoric sentence must be matched to its appropriate interpretation in a set of four pictures?

(2) How do various groups of patients compare with one another on this task? In particular, what are the effects of unilateral brain lesions (left vs. right hemisphere patients); different forms of aphasia (anterior vs. posterior aphasia); focal vs. diffuse disorders (unilateral lesions vs. dementias due to bilateral brain disease)?

(3) How do the various groups of patients perform on a task requiring a verbal explication of a metaphor? What is the relationship within each group between performance on this linguistic task and the above pictorial measure of metaphoric competence?

(4) What qualitative differences emerge among various groups of patients? In particular, do profiles of errors, reasoning, and overall reactions to the stimuli differentiate among the various subject populations?

## DISCUSSION

This study calls into question the oversimplified characterizations of the two hemispheres introduced earlier.

On the one hand, the difficulties displayed by the patients with right-hemisphere lesions in matching metaphors with pictures, along with their un-questioned acceptance of the literal depictions and their occasional reluctance to accept the verbal metaphors, indicate that an intact left hemisphere does not of itself ensure adequate comprehension of all linguistic messages.

On the other hand, the difficulties displayed by the patients with lesions of the left hemisphere in offering verbal explications of the metaphors and the re-duced performance of patients with left-posterior lesions on the pictorial test documents that an intact right hemisphere does not guarantee adequate aesthetic sensitivity.

This double dissociation of performances on the two metaphor tests clarifies the contributions made by each hemisphere to linguistic and aesthetic function-ing. In particular, the left hemisphere appears crucial for full appreciation of the denotation of words; the right hemisphere is necessary for the acceptance of con-notative language, the detection of absurd or humorous content, and the map-ping of figurative language onto situations in which it is appropriate. Apparently only a complementary interaction of the two hemispheres permits fully adequate performance in both the linguistic and the aesthetic realms.

The difficulties exhibited by left-hemisphere patients in offering a verbal ex-plication are unremarkable in view of these patients' demonstrated difficulties in verbal output, and particularly in producing abstract lexical items. The unex-pected dissociation in the right-hemisphere patients between metaphoric verbal explications and literal picture selections deserves further comment.

To begin with, the competent performance of right-hemisphere patients on the verbal condition invalidates the assertion that they are insensitive to metaphor or that they, like demented patients, have regressed to a more primitive level of functioning. However, their tendency to respond literally on the pictorial task must be accounted for.

Several alternative explanations suggest themselves. It is possible, first of all, that the anomalous performance of these subjects is explicable in terms of size of lesion. Even in a study like the present one, where all patients conforming to criteria of selection are included, right-hemisphere lesions may on the average be larger than left-hemisphere lesions: the latter, yielding aphasic symptoms, come to clinical notice more quickly.

On the other hand, there is evidence that abilities are less focally organized in the right hemisphere. Thus, even if the right-hemisphere lesions were some-what larger, this fact need not in and of itself undermine the present findings: Given possible differences in hemispheric organizations, it is not clear how best to "match" lesions in the two hemispheres. Moreover, the fact that patients with global lesions of the left hemisphere (involving both anterior and posterior regions) exhibited the same pattern of response as patients with more restricted damage to the left hemisphere suggests that the difficulty experienced by patients with lesions of the right hemisphere is more likely to be a consequence of the site of the lesion than of its size.

It is also possible that the behavior of patients with right-sided lesions is due to visuospatial deficits. Perhaps these subjects either neglected the pictures on the right side, or had difficulty "reading" the pictures per se. This first possibility, however, loses plausibility in light of the fact that these patients clearly examined each picture before responding, often commenting upon and touching each one before making a selection. The second possibility, that right-hemisphere performance was due to difficulty in interpreting the pictures, must also be questioned in light of the fact that the patients did not respond with equal likelihood to all four pictures. Indeed, even if these patients exhibited some visuospatial problems, the consistency with which many of them responded (literal first, metaphoric second) belies any simple interpretation such as difficulty in reading pictures per se. Such an account could neither explain why these patients favored the literal depictions nor explain why—unlike their left-hemisphere-damaged counterparts—they failed to find them amusing.

It seems parsimonious to attribute to the right-hemisphere patients a qualitatively different mode of appreciation of metaphor, one that reflects their overall orientation to experience. In general, aphasic patients appear to be well oriented, reasonably sensitive to connotation, able to react appropriately in a variety of situations, and quite capable of using nonlinguistic cues; in fact, the latter capacity may aid them in eliminating the implausible pictures.

In contrast, while typically relevant and accurate in their linguistic responses, right-hemisphere patients often appear insensitive to their surroundings and inappropriate in their emotional reactions. Moreover, right-hemisphere patients are not infrequently witty or metaphoric, but these characteristics emerge with undue frequency in situations where they are manifestly inappropriate.

Should these impressions be valid, they would predict the findings that have in fact been obtained: a capacity, when so directed, to offer a suitable verbal explication, but a correlative inability to identify the situation in which such a remark would be apt and an unawareness that certain pictorial depictions are in and of themselves bizarre. To the extent that such sensitivity to nuance and situation is deemed crucial in artistic matters, the claim that the right hemisphere is critical for aesthetic processing receives at least partial support.

## IN SUM

The results challenge both the "linguistic" and the "aesthetic" characterizations of cerebral lateralization. An alternative account is proposed in which both hemispheres contribute in characteristic ways to metaphoric competence and, more generally, to linguistic and aesthetic sensitivity.

## ACKNOWLEDGMENT

We thank Norman Geschwind and Edwin Weinstein for their insights about patients with right-hemisphere lesions.

## REFERENCE

Winner, E., & Gardner, H. (1977). The comprehension of metaphor in brain-damaged patients. *Brain*, *100*(4), 717–729.

# The Stories of the Right Hemisphere
## Missing the Point

I had never thought of myself as an expert on motivation. Indeed, I have long chuckled at the remark once made to me by my colleague Mihaly Csikszentmihalyi. "Howard," he quipped, "cognition is the easy part, it's motivation that's the formidable challenge."

While not pretending to be a scholar of motivation, I used this invitation as an opportunity to reflect on what we had learned from nearly 2 decades of research with brain-damaged patients. And I placed this under the umbrella of "the stories of the right hemisphere." With this pun, I sought both to summarize the role of the right hemisphere in understanding narratives, and to put together our own narrative about the major cognitive burdens of one half of the brain.

### THE STORIES OF THE RIGHT HEMISPHERE

As a student in developmental psychology with a special interest in artistic development, my scholarly life changed decisively in the late 1960s. Steeped in the writings of Piaget, while almost totally ignorant of work on the brain, I happened to meet the eminent neurologist Norman Geschwind, then at the height of his creative powers (see Essay 4).

Norman Geschwind convinced me, and my colleagues at Harvard Project Zero, that there existed a world that cried out for investigation. That world, which he did so much to elucidate, was the world of the brain-damaged patient—the once "normal" individual who had the misfortune of suffering damage to his or her brain and who now exhibited an unexpected profile of severe intellectual deficits and islands of preserved competence. Not only were the cases described by Geschwind fascinating in themselves (in a way that we have come to associate with the writings of Oliver Sacks), but they also held that promise of illuminating the nature of artistry. Geschwind described instances of famous artists—like the French composer Maurice Ravel and the French painter André Derain—whose work had been altered as a consequence of injury to the brain. In broader terms, he held forth the hope that just as studies of aphasia have helped to explain normal language, sustained study of such brain-injured musicians and painters could help to reveal the nature of artistic capacities.

When I first arrived at the Veterans Administration Hospital, it appeared as though my wildest fantasies would be realized. The first patient I saw at rounds was described as a singer, the second as a painter. Alas, these patients did not live up to their advance billing: The singer had sung only in his high school chorus, while the painter had restricted his talents to houses and commercial buildings!

While my immediate research questions seemed unlikely to be answered, I found myself increasingly intrigued by the phenomena of the neuropsychological ward. Before long, I was writing standard psychological papers on aphasia, alexia, agnosia, and other clinical conditions (see Essays 13–16). Interesting cases occasionally came to my attention, but the supply of artists who made their way to the wards of the Boston VA was meager indeed.

After a while, however, I realized that I could nonetheless continue my research interest in the arts. Instead of investigating the decomposition of highly complex skills in expert artists, I could look at the kinds of artistic and arts-related abilities possessed by ordinary individuals at this hospital—in this case, in the middle 1970s, nearly all of them male veterans of World War II. In working with this population, I could see how these skills and abilities were or were not affected by various kinds of brain injury, especially strokes. And thus, over the years, I probed the capacities of brain-injured patients to draw, to sing, to tell stories, and to carry out other kinds of art-related activities.

Most of my studies dealt in some way with language and language disturbances. In such studies, it was common to employ a brain-damaged control group. Since most aphasics have sustained damage to the left hemisphere of the brain, my research typically used as a control group individuals who had sustained damage to comparable areas in the right hemisphere. And indeed, when it came to tasks like syntactic processing, detecting word meaning, reading, writing, and naming, right-hemisphere-damaged (RHD) patients responded much more like normal individuals than like individuals who had sustained pathology in the left hemisphere of the brain.

However, this was not always the case. My associates and I began to notice that sometimes the RHD patients behaved in ways that were quite anomalous. For instance, in a study of sensitivity to jokes and cartoons, RHD patients exhibited rank orderings of understanding and of appreciation that differed more from those of normal controls than did the performances of left-hemisphere-injured (LHD) patients. In a study of sensitivity to word connotation, RHD patients resisted participating in the experimental paradigm and, when they did, proved far better at matching words to their denotations than to their connotations; LHD patients exhibited no such dissociations. And in a study of story production, the RHD patients not only had difficulty in following the fate of specific characters, but also often seemed to miss the point of the story.

At about the time that I was beginning to think that the behavior of the RHD patient might be bizarre in the linguistic or emotional realm, a well-publicized political event reinforced my speculations. Supreme Court Justice William O. Douglas suffered a right-hemisphere stroke on New Year's Eve in

1974. As is often the case in such matters, the press and the Court spokesperson tended to downplay the severity of the stroke. Yet, as I read the reports from Douglas's physicians and observed him occasionally on television news programs, I found his conformity to the emerging "RHD patient profile" to be quite telling.

Douglas could speak well enough, but a good bit of what he said did not make much sense. It seemed on target for a while, but then veered away. He sought to make light of his own injury even as he failed to appreciate the underlying intent of questions that were posed to him. Later, it turned out that Douglas caused enormous difficulty for his "brethren" on the Court. Not only did he fail to realize his own deficits and instead insist that he was fit to serve, but his understanding of new legal cases and his ability to draw appropriately on his years of experience were clearly impaired.

When encountering a new phenomenon, one is tempted to determine its limits. Thus, in the following years, my colleagues and I probed and documented a whole ensemble of difficulties exhibited by RHD patients in language and language-related tasks. In addition to the problems with connotation and humor noted above, RHD patients turn out to have difficulty appreciating figurative language—metaphors as well as sarcastic and ironic statements—comprehending indirect speech acts (e.g., requests made by implication), organizing sentences into a coherent paragraph, drawing inferences rather than responding to direct assertion from one sentence to another, monitoring and correcting deviant conversations, and understanding several other spheres as well.

For a while it looked as though RHD patients had difficulty with almost any aspect of language beyond simple denotative reference at the sentence level. Moreover, my colleagues and I documented a special complex of problems with stories. Once a story went beyond a familiar, canonical member of the genre, RHD patients proved unable to handle these texts. They could not reproduce coherent texts on their own, and when given a text written by someone else, they exhibited special problems with determining reference, coping with surprise, and discerning underlying morals.

Curious about whether these difficulties pertained to all kinds of stories, my colleagues and I sought to examine narratives that contained specific kinds of content. These studies were revealing: If the content could be expressed entirely in words and there was no apparent need to construct some kind of visual model of the action, patients could handle the texts reasonably well. Once a story seemed to require some kind of visual imaging, however, RHD patients were not successful. We also found that if emotional content could be captured in words alone, patients understood a scene that was discussed; but if patients had to infer the emotions or impute them to characters on the basis of situational or personality traits, then they were often at a loss to make sense of the story. Finally, they experienced great difficulty with stories that did not confirm to already-mastered genres, such as a script or joke format. If a story contained unfamiliar materials, subjects would either treat them as familiar or ignore them altogether.

Considering the gamut of clinical observations and experimental findings, I propose that, in the processing of language, the right hemisphere plays a specific role: that of *monitoring the message at the level of emotional continuity and discontinuity*. It is the province of the right hemisphere to encode linguistic messages *not* in terms of the semantic content but rather in terms of the *emotional weight* attached to actions, to descriptions, and to the relations among them.

Let me provide some background to this claim. When an individual processes any kind of extended discourse, such as a story, a joke, a vignette, and so on, the individual must be able to follow syntax, phonology, and literal word meaning. Such monitoring is handled comfortably by the left, or linguistic, hemisphere. Moreover, to be able to follow the discourse, the individual must be able to appreciate the genre in which it is encoded: a short story, a shaggy dog joke, a legal deposition, and so forth.

Contrary to what one might have supposed, RHD patients seem capable of appreciating instances of a genre, such as a script. Moreover, they can both create and appreciate canonical instances of that genre. Thus, to the extent that an extended passage of prose follows the usual script, such patients are able to perform quite normally. In this context, it is worth noting that Justice Douglas, while unable to render judgments about new cases, was able to characterize old cases with considerable acuity, going well beyond strictly memorized utterances.

Where, then, does the RHD patient exhibit deficits? I contend that, under normal situations, individuals monitor discourse in parallel in two distinct ways: in terms of *overt content* (in which the left hemisphere performs a serviceable job on its own) and in terms of the *conformities to and deviations from normal expectations* within the genre. When there is a deviation of some sort from expected content or ordering, the right hemisphere registers this discrepancy as an emotional jag or anomaly—with accompanying surprise, laughter, or grimace of disappointment. In works of art, such deviations can be quite complex, as an ensemble of emotions comes to accompany a particular character, scene, or tableau and changes in accord with the fate of that element. A work of art may harbor its own emotional landscape, with a characteristic anatomy and physiology.

What happens to the individual who—wholly or in part—loses the ability to follow discourse at this emotional level?

To begin with, the individual may note a deviation from expectation but not know what to make of it. In the absence of an appropriate interpretation or categorization, it is difficult to store the bit of information. The individual instead reports something as being anomalous or, more likely, reverts to the canonical version, the only one that he or she can still deal with. The more severe the impairment, the less likely that the deviation will be detected and registered, let alone processed appropriately. Such appropriate encoding, storage, and processing seems to require intact emotional monitoring.

How can this characterization account for the data on the comprehension of discourse? As noted, we find in our studies that RHD patients are able to process and remember script and script-like entities. When it comes to jokes or stories with a surprise ending, they are able to appreciate that something is noncanonical. And yet, when asked to reconstruct a stimulus item or continue it in some way, such patients are unable to do so. When dealing with conversation, they are unable to appreciate the reasons someone might have violated a canon of discourse, and thus they end up acquiescing to any kind of conversational move. By the same token, when attempting to understand indirect discourse, the individual does not appreciate the reasons underlying certain requests or statements and again falls back on the most canonical explanation.

One may ask how this pattern of behavior relates to patients' own linguistic production. This is a difficult question. Right-hemisphere-damaged patients usually speak in a comprehensible way, yet they also tend to go off on tangents, to be inappropriately jocular, and to embrace metaphors that may be inappropriate or even nonsensical.

I believe that this strange manner of speaking may be related to the putative difficulty I have hypothesized. It seems to me that RHD patients, in a manner somewhat reminiscent of schizophrenic speakers, are susceptible to intoxication by the associations between one phrase or word and another. If they are consciously aware of them, normal speakers censor such associations because they know that others will find them bizarre. Bereft of an emotional "shadower," lacking an emotional gyroscope to keep them on course, RHD patients are unable to appreciate the effects of a deviation, and so they continue speaking blithely in a linguistically fluent manner.

When it comes to pathology, it is often useful to see whether a reported condition can be related to one's own phenomenal experience. Let me illustrate the role that the right hemisphere apparently plays for me in contexts such as these. Not infrequently, and more so the older one gets, one begins a book or attends a movie without knowing that this particular instance of the genre has already been encountered. Sometimes, of course, one immediately recognizes the work of art, and the contents of the work emerge intact. At other times, however, the work exudes a certain familiarity and one struggles to figure out whether it has in fact been encountered previously.

In such instances, I find that emotional encoding is often more reliable than semantic coding. That is, I recognize and can anticipate the mood of the piece, including wide shifts in tone, well before I know exactly which characters will appear and what will happen to them. This experience seems to model the role of the right hemisphere—the supplying of a kind of emotional glue that helps to confer coherence on unfamiliar narrative discourses. Or, as Nelson Goodman might put it, the emotions are functioning cognitively (see Essay 3).

While the domain of linguistic regularity, deviation, and anomaly is usually deemed the special province of the arts, I submit that this territory of emotional nuance actually pervades much of our existence. Political candidates, for example, compete with one another to convey stories that are convincing to the electorate. Usually they simply manipulate classical scripts, hoping to convince the electorate that their script is the most faithful to beliefs and wishes that are already widely held. However, the more adventurous leader is able to boost the stock of stories or scripts that have not been appreciated for some time and bring them to the fore; such, I believe, were the accomplishments of such disparate leaders as Margaret Thatcher and Franklin Roosevelt. Unless listeners can detect and draw proper inferences from these deviations, they will either be missed entirely or misinterpreted (see Essays 25 and 26).

Such forces are even at work in the sciences. Those who attempt to explain human or natural phenomena begin with the accounts that are already prevalent in the literature and toy with them in various ways. Thus, in this very presentation, I began with a capsule story of my own professional development. I then shifted the narrative to an account of a set of empirical discoveries, reviewing the usual explanations that have been offered. Following this presentation, I then introduced a new line of analysis that holds promise of explaining some proportion of the disparate literature on linguistic processing in RHD patients. As one encounters this novel idea, I suggest, not only must its literal meaning be processed but also the idea must be located emotionally: that is, as something liked or disliked, something that is or is not consistent with one's belief system, something that seems promising for suggesting new experiments, or something that is limited or even already falsified. Such dual coding is not simply a luxury; rather, it contributes measurably to the comprehensibility and to the memorability of the particular explanatory account that I have introduced here.

What investigative lanes are opened? From an experimental point of view, one can investigate the generality of this hypothesized phenomenon. For example, does the right hemisphere participate in dual coding for linguistic messages only; or, as examples from film or cartoons suggest, is it also relevant for materials encoded in other systems of symbols? Are there certain kinds of emotional messages or connections that are more readily processed by such patients? For instance, our studies of story comprehension suggest that when emotional charges are very potent, RHD patients are more likely to be able to process them. The findings to which I refer suggest that the degree of emotional potency is relevant; it is possible, however, that certain emotional states (e.g., positive ones) may be more readily apprehended by RHD patients.

There are also clinical and therapeutic implications. In trying to communicate with RHD patients, it may be prudent to recognize that they are unlikely to be able to present, or to follow, a set of emotional associations. Accordingly, one must pay attention chiefly to the literal content of what they say.

Finally, to the extent that such patients can benefit from therapeutic interventions, it may be advisable to help them appreciate the emotional implications of what they are saying—either by making these explicit, by infusing contents with greater-than-normal emotional weight, or by reacting to linguistic output with specific emotional signals. For the rest of us, such excessive reactions are not necessary.

In that spirit, I would ask you to mute your reaction to this article—unless, of course, it is wholly positive . . .

# INTRODUCTION TO MULTIPLE INTELLIGENCES

In the decade following the conclusion of my doctoral studies, I was continuing down a daily research path: neuropsychology at the Veterans Administration hospital in the morning; child development in school and at the Project Zero research lab in the afternoon.

But then, one day in 1979, I learned that the Harvard Graduate School of Education had received a very large grant (then over $1 million—in today's dollars a multiple of this). The grant presented a rather grandiose assignment: *to illuminate the nature and the realization of "human potential."* At the time I quipped that that was "more of an [American] West Coast than an East Coast assignment." I was invited to co-direct the project. In that role, I assumed a grand and inviting assignment: to conduct research and then write about the nature and realization of human cognitive capacities.

I had already sought to put together what I was learning about human cognitive and symbolic development (particularly in the arts) with what I was documenting about the breakdown of such capacities under conditions of brain damage. Now, as if by magic, I was given a once-in-a-lifetime opportunity—along with assistants and a generous budget—to travel to many parts of the world; consult with major scholars across the social sciences; and canvass the considerable literature in what was then coming to be called "cognitive science"—what we have learned about the mind from psychology, linguistics, neuroscience, anthropology, computational science, and—yes—2,000+ years of philosophical reflection in the West (and now, we could add, in other regions as well; see Essay 39).

As early as 1976, having worked for several years with normal and gifted children as well as brain-damaged individuals, I had sketched out a book called *Kinds of Minds.* I had the conception, perhaps, but I lacked the tools, the tenacity, and/or the terminology to consummate that book project.

But sometime around 1980–1981, the pieces gradually came together. I was able to propose for the first time the set of ideas for which I am still best known: the claim that human beings have not one or two, but at least *seven major ways of knowing the world*. Each of these potentials is, in a sense, its own computer. And so, when we think about the human mind, we should think of it as involving seven separate and at least partially autonomous computational devices.

It's lost to posterity exactly how I came up with the name, but by the early 1980s I was speaking readily about "human intelligences," and in 1983, I published my most influential book, *Frames of Mind: The Theory of Multiple Intelligences*.

If, before reading the previous paragraph, you had known *anything* about me and my work, you would probably know this: Howard Gardner claims that the human mind contains seven (or thereabouts) separate intelligences. Most intelligence tests capture important aspects of linguistic and logical-mathematical intelligences (we might call them the "scholastic intelligences") and some tests also tap spatial intelligence. But Gardner now claims that there are several additional intelligences: in 1983, musical, bodily-kinesthetic, interpersonal, and intrapersonal intelligences; by 1994, the naturalist intelligence; and thereafter, possibly *existential* intelligence (the capacity to raise and ponder big questions) and *pedagogical* intelligence (the capacity to teach others).

Here I present the theory in a nutshell, as well as some major critiques and my responses; and, importantly, the major misunderstandings. For the educational implications of the theory, please see the companion volume, *The Essential Howard Gardner on Education*.

## REFERENCE

Gardner, H. (1983). *Frames of mind: The theory of multiple intelligences*. Basic Books.

# In a Nutshell

*I have written several books about MI theory. and many articles as well. But it is good to have a capsule view of the theory—and that was best provided in my early writings from the 1980s.*

One of the most important events in the history of psychology occurred over a century ago. As is well known, in the first decade of the 20th century, French psychologist Alfred Binet devised the first tests of human intelligence or intellect. Binet's motivation was straightforward; he wanted to be able to identify *which* children would sail through the early grades of school; *which* children were likely to encounter learning challenges; and how best to help the latter group succeed in school.

Like other Parisian fashions, these tests soon made their way across the Atlantic. Binet's American counterparts adapted the tests for use with individuals of different ages and backgrounds and for a wide range of purposes—including placement in the military or in occupational niches. They shortened the term "intelligent quotient" to IQ; and within a few decades, they devised a Scholastic Aptitude Test (SAT)—an instrument subsequently taken by millions of high school students in order to determine the college(s) for which they were suited.

Why do I consider the invention of the IQ test—and its psychometric progeny—to be so important for the line of inquiry in which I've been engaged? I think it has to do with the ideals of our own society—of Western society dating back hundreds, perhaps thousands of years, perhaps dating back to Greece—that what is *truly important* is *human rationality*. That's how we rate people—how rational they are, how well they can solve problems. We have a word called "smart." We talk about people as being smart or smarter; and of course, we talk about people being dumb or dumber. Now, the West is also very interested in quantities, and so we have invented rulers that measure how tall somebody is; and we also have charts so that if we know how tall somebody is at 3 years of age, we can predict how tall she will be when she is grown up. Just multiply toddler height by two. Now having been able to quantify physical growth, wouldn't it be great if we could also quantify mental power and mental growth? And of course, that's what has happened with the Binet-SAT enterprise over the last century.

As you may well have anticipated, I am not entirely satisfied with this state of affairs! Two quick reasons:

l. There have been efforts to create even shorter instruments. One such measure called the QT, "quick test," purports to "measure someone's intelligence in three to five minutes." From my perspective, this is simply blustering, not serious intervention.

2. Some researchers propose to eliminate the "Q and A" altogether and just look at a pattern of brain waves. While such measures have modest correlations with other tests of intelligence, they can scarcely claim to indicate an individual's intellectual process.

I also have a subtler tension with the notion of measuring intelligence. At one point, the great English humanist, Dr. Samuel Johnson, was asked, "What is intelligence?" Dr. Johnson responded, "Well, genius is an individual of large general powers, accidentally deflected in one direction or another." Dr. Johnson lived nearly two hundred years before Charles Spearman, who coined the term "$g$" for General Intelligence. This was an early claim for General Intelligence. If you had a lot of it, you could use it for anything. It was a sort of polymorphous intellect. You could be a politician, a poet, a philosopher, or a pianist. Everything was open to you. If you haven't got a lot of it, forget it. You couldn't do anything significant.

I've developed an alternative vision—one based on a radically different view of the mind, and one that yields a very different view of school. It is a *pluralistic view of mind*, recognizing many different and discrete facets of cognition, acknowledging that people have different cognitive strengths and contrasting cognitive styles.

To introduce this new point of view, let us undertake the following "thought experiment." Suspend the usual judgment of what constitutes intelligence, and let your thoughts run freely over the capabilities of humans—perhaps those that would be picked out by the proverbial visitor from Mars. In this exercise, you are drawn to the brilliant chess player, the world-class violinist, and the champion athlete; such outstanding performers deserve special consideration. Following through on this experiment, a quite different view of intelligence emerges. Are the chess player, violinist, and athlete "intelligent" in these pursuits? If they are, then why do our tests of "intelligence" fail to identify them? If they are *not* "intelligent," what allows them to achieve such astounding feats? In general, why does the contemporary construct of "intelligence" fail to take into account large areas of human endeavor?

To approach these questions, I introduced the theory of multiple intelligences (MI). As the pluralistic name signals, I believe that human cognitive competence is better described in terms of a set of abilities, talents, or mental skills, which I call "intelligences." All normal individuals possess each of these skills to some extent; individuals differ in the degree of skill and in the nature of their combination. I believe that this theory of intelligence may be more humane and more veridical than alternative views of intelligence and that it more adequately reflects the data of human "intelligent" behavior. Such a theory has important educational implications. (See Essays 8–15 in *The Essential Howard Gardner on Education*.)

## WHAT CONSTITUTES AN INTELLIGENCE?

The question of the optimal definition of intelligence looms large in this inquiry. And it is here that the theory of multiple intelligences begins to diverge from traditional points of view. In the classic psychometric view, intelligence is defined operationally as the ability to answer items on tests of intelligence. The inference from the test scores to some underlying ability is supported by statistical techniques. These techniques compare responses of subjects at different ages; the apparent correlation of these test scores across ages and across different tests corroborates the notion that the general faculty of intelligence, called *g* in short, does not change much with age, training, or experience. It is an inborn attribute or faculty of the individual.

Multiple intelligences theory, on the other hand, pluralizes the traditional concept. An intelligence is a *computational capacity*—a capacity to process a certain kind of information—that is a component of human biology and human psychology. Humans have certain kinds of intelligences, whereas rats, birds, and computers foreground other kinds of computational capacities.

An intelligence entails the ability to solve problems or fashion products that are of consequence in a particular cultural setting or community. The problem-solving skill allows one to approach a situation in which a goal is to be obtained and to locate the appropriate route to that goal. The creation of a cultural product allows one to capture and transmit knowledge or to express one's conclusions, beliefs, or feelings. The problems to be solved range from creating an end for a story to anticipating a mating move in chess to repairing a quilt. Products range from scientific theories to musical compositions to successful political campaigns.

MI theory is framed in light of the biological origins of each problem-solving skill. Only those skills that are universal to the human species are considered (again, we differ from rats, birds, or computers). Even so, the biological proclivity to participate in a particular form of problem-solving must also be coupled with the cultural nurturing of that domain. For example, language, a universal skill, may manifest itself particularly as writing in one culture, as oratory in another culture, and as the secret code composed of anagrams in a third.

Given the desire of selecting intelligences that are rooted in biology, and that are valued in one or more cultural settings, how does one actually identify an "intelligence"? In coming up with the list, I reviewed evidence from several different sources:

- knowledge about normal development and development in gifted individuals;
- information about the breakdown of cognitive skills under conditions of brain damage;
- studies of exceptional populations, including prodigies, savants of various sorts, and autistic children;
- data about the evolution of cognition over the millennia;

- cross-cultural accounts of cognition;
- psychometric studies, including examinations of correlations among tests; and
- psychological training studies, particularly measures of transfer and generalization across tasks.

Only those candidate intelligences that satisfied all or a healthy majority of the criteria were selected as bona fide intelligences. A more complete discussion of each of these criteria for an "intelligence," and the intelligences that were initially identified, is found in *Frames of Mind* (Gardner, 1983).

In addition to satisfying the aforementioned criteria, each intelligence must have an identifiable core operation or set of operations. As a neurally based computational system, each intelligence is activated or "triggered" by certain kinds of internally or externally presented information. For example, one *core* of musical intelligence is the sensitivity to pitch relations, whereas one *core* of linguistic intelligence is the sensitivity to phonological features.

An intelligence must also be susceptible to *encoding in a symbol system*—a culturally contrived system of meaning that captures and conveys important forms of information. Language, picturing, and mathematics are but three nearly worldwide symbol systems that are necessary for human survival and productivity.

The relationship of a candidate intelligence to a human symbol system is no accident. In fact, the existence of a core computational capacity anticipates the actual or potential creation of a symbol system that exploits that capacity. While it may be possible for an intelligence to develop without an accompanying symbol system, a primary characteristic of human intelligence may well be its gravitation toward such an embodiment.

## THE ORIGINAL SET OF INTELLIGENCES

I turn now to a brief consideration of each of the intelligences that were proposed in the early 1980s. I begin each sketch with a thumbnail biography of a person who demonstrates an unusual facility with that intelligence. Although each biography illustrates a particular intelligence, I do not wish to imply that in adulthood intelligences operate in isolation. Indeed, except in individuals with neuro-atypical profiles, intelligences work in concert, and any sophisticated adult role will likely involve a melding of several of them.

*Musical Intelligence.* When he was 3 years old, Yehudi Menuhin was smuggled into the San Francisco Orchestra concerts by his parents. The sound of Louis Persinger's violin so entranced the youngster that he insisted on a violin for his birthday and Louis Persinger as his teacher. He got both. By the time he was 10 years old, Menuhin was an international performer (Menuhin & Davis, 1979).

Menuhin's musical intelligence manifested itself even before he had touched a violin or received any musical training. His powerful reaction to that particular sound and his rapid progress on the instrument suggest that he was biologically prepared in some way for that endeavor. In this way, evidence from child prodigies supports the claim that there is a biological link to a particular intelligence. Other special populations, such as autistic children who can play a musical instrument beautifully but who cannot otherwise express themselves, underscore the independence of musical intelligence.

A brief consideration of the evidence suggests that musical skill passes the other tests for an intelligence. For example, certain parts of the brain play important roles in perception and production of music. These areas are characteristically located in the right hemisphere, although musical skill is not as clearly "localized," or located in a specifiable area, as natural language. Although the particular susceptibility of musical ability to brain damage depends on the degree of training and other individual differences, there is clear evidence for "amusia," or loss of musical ability.

*Bodily-Kinesthetic Intelligence.* Fifteen-year-old Babe Ruth was playing catcher one game when his team was taking a "terrific beating." Ruth "burst out laughing" and criticized the pitcher loudly. Brother Mathias, the coach, called out, "All right, George, YOU pitch!" Ruth was stunned and nervous: "I never pitched in my life . . . I can't pitch." The moment was transformative, as Ruth recalls in his autobiography: "Yet, as I took the position, I felt a strange relationship between myself and that pitcher's mound. I felt, somehow, as if I had been born out there and that this was a kind of home for me" (Ruth & Considine, 1948, p. 17). As sports history shows, he went on to become a great major league pitcher (and, of course, attained legendary status as a hitter).

Like Menuhin, Babe Ruth was a prodigy who recognized his "instrument" upon his first exposure to it. This recognition occurred in advance of formal training.

Control of bodily movement is, of course, localized in the motor cortex, with each hemisphere dominant or controlling bodily movements on the contralateral side. The ability to perform movements when directed to do so can be impaired even in individuals who can perform the same movements reflexively or on a nonvoluntary basis. The existence of specific apraxia constitutes one line of evidence for a bodily-kinesthetic intelligence.

The evolution of specialized body movements is of obvious advantage to the species, and in humans this adaptation is extended through the use of tools. Body movement undergoes a clearly defined developmental schedule in children; there is little doubt of its universality across cultures. Thus it appears that bodily-kinesthetic "knowledge" satisfies many of the criteria for an intelligence.

The consideration of bodily-kinesthetic knowledge as "problem-solving" may be less intuitive. Certainly carrying out a mime sequence or hitting a tennis ball is not solving a mathematical equation. And yet, the ability to use one's body to express an emotion (as in a dance), to play a game (as in a sport), or to create a new product (as in devising an invention) is evidence of the cognitive features of body usage.

*Logical-Mathematical Intelligence.* In 1983 Barbara McClintock won the Nobel Prize in Medicine or Physiology for her work in microbiology. Her intellectual powers of deduction and observation illustrate one form of logical-mathematical intelligence that is often labeled "scientific thinking." One incident is particularly illuminating. As a researcher at Cornell in the 1920s, McClintock was faced one day with a problem: While theory predicted 50% pollen sterility in corn, her research assistant (in the "field") was finding plants that were only 25% to 30% sterile. Disturbed by this discrepancy, McClintock left the cornfield and returned to her office where she sat for half an hour, thinking:

> Suddenly I jumped up and ran back to the [corn] field. At the top of the field [the others were still at the bottom] I shouted, "Eureka, I have it! I know what the 30% sterility is!" . . . They asked me to prove it. I sat down with a paper bag and a pencil and I started from scratch, which I had not done at all in my laboratory. It had all been done so fast; the answer came and I ran. Now I worked it out step by step—it was an intricate series of steps—and I came out with [the same result]. [They] looked at the material and it was exactly as I'd said it was; it worked out exactly as I had diagrammed it. Now, why did I know, without having done it on paper? Why was I so sure? (E. F. Keller, 1983, p. 104)

This anecdote illustrates two essential facts of the logical-mathematical intelligence. First, in the gifted individual, the process of problem-solving is often remarkably rapid—the successful scientist copes with many variables at once and creates numerous hypotheses that are each evaluated and then accepted or rejected in turn.

The anecdote also underscores the nonverbal nature of the intelligence. A solution to a problem can be constructed before it is articulated. In fact, the solution process may be totally invisible, even to the problem-solver. This phenomenon need not imply, however, that discoveries of this sort—the familiar "aha!"—are mysterious, intuitive, or unpredictable. The fact that it happens more frequently to some people (e.g., Nobel Prize winners) suggests the opposite. We interpret this as the work of the logical-mathematical intelligence.

Along with the companion skill of language, logical-mathematical reasoning provides the principal basis for IQ tests. This form of intelligence has been thoroughly investigated by traditional psychologists, and it is the archetype of "raw intelligence," or the problem-solving faculty that purportedly cuts across domains. It is perhaps ironic, then, that the actual mechanism by which one arrives at a solution to a logical-mathematical problem is not as yet completely understood—and the processes involved in leaps like those described by McClintock remain mysterious.

*Linguistic Intelligence.* At the age of 10, T. S. Eliot created a magazine called *Fireside* to which he was the sole contributor. In a 3-day period during his winter vacation, he created eight complete issues. Each one included poems, adventure stories, a gossip column, and humor. Some of this material survives, and it displays the talent of the poet.

As with the logical intelligence, calling linguistic skill an "intelligence" is consistent with the stance of traditional psychology. Linguistic intelligence also passes our empirical tests. For instance, a specific area of the brain, called Broca's area, is responsible for the production of grammatical sentences. A person with damage to this area can understand words and sentences quite well but has difficulty putting words together in anything other than the simplest of sentences. At the same time, other thought processes may be entirely unaffected.

The gift of language is universal, and its rapid and unproblematic development in most children is strikingly constant across cultures. Even in deaf populations where a manual sign language is not explicitly taught, children will often "invent" their own manual language and use it surreptitiously! We thus see how an intelligence may operate independently of a specific input modality or output channel.

*Spatial Intelligence.* Navigation around the Caroline Islands in the South Seas is accomplished by native sailors without instruments. The position of the stars as viewed from various islands, the weather patterns, and water color are the principal signposts. Each journey is broken into a series of segments, and the navigator learns the position of the stars within each of these segments. During the actual trip the navigator must envision mentally a reference island as it passes under a particular star. From that he computes the number of segments completed, the proportion of the trip remaining, and any corrections in heading that are required. The navigator cannot see the islands as he sails along; instead he maps their locations in his mental "picture" of the journey (see Gladwin, 1970).

Spatial problem-solving is required for navigation and in the use of the notational system of maps. Other kinds of spatial problem-solving are brought to bear in visualizing an object seen from a different angle and in playing chess. The visual arts also employ this intelligence in the use of space.

Evidence from brain research is clear and persuasive. Just as the middle regions of the left cerebral cortex have, over the course of evolution, been selected as the site of linguistic processing in right-handed persons, the posterior regions of the right cerebral cortex prove most crucial for spatial processing. Damage to these regions causes impairment of the ability to find one's way around a site, to recognize faces or scenes, or to notice fine details.

Blind populations provide an illustration of the distinction between spatial intelligence and visual perception. A blind person can recognize shapes by an indirect method: running a hand along the object translates into length of time of movement, which in turn is translated into the size of the object. For the blind person, the perceptual system of the tactile modality parallels the visual modality in the seeing person. The analogy between the spatial reasoning of the blind and the linguistic reasoning of the deaf is notable.

*Interpersonal Intelligence.* With little formal training in special education and nearly blind herself, Anne Sullivan began the intimidating task of instructing a blind and deaf 7-year-old, Helen Keller. Sullivan's efforts at communication were complicated by the child's emotional struggle with the world around her. At their first meal together, this scene occurred:

> Annie did not allow Helen to put her hand into Annie's plate and take what she wanted, as she had been accustomed to do with her family. It became a test of wills—hand thrust into plate, hand firmly put aside. The family, much upset, left the dining room. Annie locked the door and proceeded to eat her breakfast while Helen lay on the floor kicking and screaming, pushing and pulling at Annie's chair. [After half an hour] Helen went around the table looking for her family. She discovered no one else was there and that bewildered her. Finally, she sat down and began to eat her breakfast, but with her hands. Annie gave her a spoon. Down on the floor it clattered, and the contest of wills began anew. (Lash, 1980, p. 52)

Anne Sullivan sensitively responded to the child's behavior. She wrote home: "The greatest problem I shall have to solve is how to discipline and control her without breaking her spirit. I shall go rather slowly at first and try to win her love." In fact, the first "miracle" occurred 2 weeks later, well before the famous incident at the pump house. Annie had taken Helen to a small cottage near the family's house, where they could live alone. After 7 days together, Helen's personality suddenly underwent a change—the therapy had worked: "My heart is singing with joy this morning. A miracle has happened! The wild little creature of two weeks ago has been transformed into a gentle child" (Lash, 1980, p. 54).

It was just 2 weeks after this that the first breakthrough in Helen's grasp of language occurred, and from that point on, she progressed with incredible speed. The key to the miracle of language was Anne Sullivan's insight into the person of Helen Keller.

Interpersonal intelligence builds on a core capacity to notice distinctions among others—in particular, contrasts in their moods, temperaments, motivations, and intentions. In more advanced forms, this intelligence permits a skilled adult to read the intentions and desires of others, even when they have been hidden. This skill appears in a highly sophisticated form in religious or political leaders, salespersons, marketers, teachers, therapists, and parents. The Helen Keller–Anne Sullivan story suggests that this interpersonal intelligence does not depend on language. All indices in brain research suggest that the frontal lobes play a prominent role in interpersonal knowledge. Damage in this area can cause profound personality changes while leaving other forms of problem-solving unharmed—after such an injury, a person is often not the "same person."

Alzheimer's disease, a form of presenile dementia, appears to attack posterior brain zones with a special ferocity, leaving spatial, logical, and linguistic computations severely impaired. Yet Alzheimer's patients will often remain well groomed, socially proper, and continually apologetic for their errors. In contrast, Pick's disease (frontotemporal dementia), another variety of presenile dementia that is localized in more frontal regions of the cortex, entails a rapid loss of social graces.

*Intrapersonal Intelligence.* In an essay called "A Sketch of the Past," written almost as a diary entry, Virginia Woolf discusses the "cotton wool of existence"—the various mundane events of life. She contrasts this "cotton wool" with three specific and poignant memories from her childhood: a fight with her brother, seeing a particular flower in the garden, and hearing of the suicide of a past visitor:

> These are three instances of exceptional moments. I often tell them over, or rather they come to the surface unexpectedly. But now for the first time I have written them down, and I realize something that I have never realized before. Two of these moments ended in a state of despair. The other ended, on the contrary, in a state of satisfaction.
>
> The sense of horror (in hearing of the suicide) held me powerless. But in the case of the flower, I found a reason; and was thus able to deal with the sensation. I was not powerless.
>
> Though I still have the peculiarity that I receive these sudden shocks, they are now always welcome; after the first surprise, I always feel instantly that they are particularly valuable. And so I go on to suppose that the shock-receiving capacity is what makes me a writer. I hazard the explanation that a shock is at once in my case followed by the desire to explain it. I feel that I have had a blow; but it is not, as I thought as a child, simply a blow from an enemy hidden behind the cotton wool of daily life; it is or will become a revelation of some order; it is a token of some real thing behind appearances; and I make it real by putting it into words. (Woolf, 1985, pp. 69–70)

This quotation vividly illustrates the intrapersonal intelligence—knowledge of the internal aspects of a person: access to one's own feeling life, one's range of emotions, the capacity to effect discriminations among these emotions and eventually to label them and to draw upon them as a means of understanding and guiding one's own behavior. A person with good intrapersonal intelligence has a viable and effective model of himself or herself—one that would be consistent with a description constructed by careful observers who know that person intimately. Since this intelligence is the most private, it requires evidence from language, music, or some other more expressive form of intelligence if the observer is to detect it at work. In the above quotation, for example, linguistic intelligence serves as a medium in which to observe intrapersonal knowledge in operation.

We see the familiar criteria at work in the intrapersonal intelligence. As with the interpersonal intelligence, the frontal lobes play a central role in personality change. Injury to the lower area of the frontal lobes is likely to produce irritability or euphoria, while injury to the higher regions is more likely to produce indifference, listlessness, slowness, and apathy—a kind of depressive personality. In such "frontal-lobe" individuals, the other cognitive functions often remain preserved.

In contrast, among aphasics who have recovered sufficiently to describe their experiences, we find consistent testimony: While there may have been a diminution of general alertness and considerable depression about the condition, the individual in no way felt himself to be a different person. He recognized his own needs, wants, and desires and tried as best he could to achieve them.

The autistic child is a prototypical example of an individual with impaired intrapersonal intelligence; indeed, the child may not even be able to refer to himself. At the same time, such children may exhibit remarkable abilities in the musical, computational, spatial, mechanical, and/or other nonpersonal realms.

In sum, then, both interpersonal and intrapersonal faculties pass the tests of an intelligence. They both feature problem-solving capacities with significance for the individual and the species. Interpersonal intelligence allows one to understand and work with others. Intrapersonal intelligence allows one to understand and work with oneself. In the individual's sense of self, one encounters a melding of interpersonal and intrapersonal components. Indeed, the sense of self emerges as one of the most marvelous of human inventions—a symbol that represents all kinds of information about a person and that is at the same time an invention that all individuals construct for themselves.

## NEWLY IDENTIFIED INTELLIGENCES

For the first 10 years after I proposed the theory of multiple intelligences, I resisted any temptation to alter the theory. Many individuals proposed candidate intelligences—humor intelligence, cooking intelligence, sexual intelligence. One of my students quipped that I would never recognize those intelligences because I lacked them myself.

Two events impelled me to consider additional intelligences. Once I spoke about the theory to several historians of science. After the conclusion of my talk, a short, elderly man approached and said, "You will never explain Charles Darwin with the set of intelligences that you proposed." The commentator was none other than Ernst Mayr, probably Darwin's successor as the most important 20th-century authority on evolution.

The other event was the frequent assertion that there was a spiritual intelligence, and the occasional assertion that I had identified a spiritual intelligence. In fact, neither statement was true. But these experiences motivated me to consider whether there is evidence for either a naturalist or a spiritual intelligence.

This inquiry led to very different conclusions. In the first case, the evidence for the existence of a naturalist intelligence is surprisingly persuasive. Human beings like biologists Charles Darwin and E. O. Wilson and ornithologists like John James Audubon and Roger Tory Peterson excel at the capacity to distinguish one species from another. An individual with a high degree of naturalist intelligence is keenly aware of how to identify the diverse, plants, animals, mountains, and cloud configurations in her ecological niche. While we tend to think of these capacities as visual, the recognition of birdsong or whale calls entails auditory perception. The Dutch naturalist Geerat Vermeij, who is blind, depends on his sense of touch.

On the eight criteria of an intelligence, the naturalist intelligence scores well. There are the core capacities to recognize instances as members of a species; the evolutionary history where survival often depends on recognizing conspecifics and on avoiding predators; and young children easily make distinctions in the naturalist world—indeed, some 5-year-olds are better at distinguishing among dinosaur species than are their parents or grandparents.

When one assumes the cultural or brain lenses, interesting phenomena emerge. Nowadays, few persons in the developed world are directly dependent on naturalist intelligence. We simply go to the grocery store or order groceries on the phone or via the Internet. And yet I suggest that our entire consumer culture is based on the naturalist intelligence. Those are the capacities on which we draw when we are drawn to one car rather than another, or when we select for purchase one pair of sneakers or gloves rather than another.

The study of brain damage provides intriguing evidence of individuals who are able to recognize and name inanimate objects but who lose the capacity to identify living things; less often, one encounters the opposite pattern, where individuals are able to recognize and name animate entities but fail with artificial (man-made) objects. It is probably the case that these capacities entail different perceptual mechanisms (Euclidean geometry operates in the world of artifacts but not in the world of nature) and different experiential bases (we operate on inanimate objects and tools in ways quite different from the ways that we interact with living beings).

My review of the evidence on *spirituality* proved less straightforward. Individuals have very strong views on religion and spirituality, particularly in the contemporary United States. For many people, experiences of the spirit are the most important ones; they assume that a spiritual intelligence not only exists but represents the highest achievement of human beings. Still others, and particularly those of a scientific bent, cannot take seriously any discussion of the spirit or the soul; it smacks of mysticism. And they may be deeply skeptical about God and religion—especially so within the academy. Asked why I did not instantly endorse a spiritual or religious intelligence, I once quipped, "If I did so, it would please my friends—but it would please my enemies even more!"

Quips are no substitute for scholarship. I devoted the better part of a year to reviewing the evidence for and against a spiritual intelligence. I concluded that at least two facets of spirituality were quite remote from my conception of an intelligence. First, I do not believe that an intelligence should be confounded with an individual's phenomenological experience. For most observers, spirituality entails a certain set of visceral reactions—for example, a feeling that one is in touch with a higher being or "at one" with the world. Such feelings may be fine, but I do not see them as valid indicators of an intelligence. A person with a high degree of mathematical intelligence may undergo feelings of "flow" when she solves a difficult problem. But she is equally mathematically intelligent even if she reports no such phenomenological reaction.

Second, for many individuals, spirituality is indissociable from a belief in religion/God generally, or even from allegiance to a particular faith or sect. "Only a real Jew/Catholic/Muslim/Protestant is a spiritual being" is the explicit or implicit message. This requirement makes me uncomfortable and takes us far from the initial set of criteria for an intelligence.

But if a spiritual intelligence does not qualify on my criteria, one facet of spirituality seems a promising candidate. I call it *existential intelligence*—sometimes described as "the intelligence of big questions." This candidate intelligence is based on the human proclivity to ponder the most fundamental questions of existence: Why do we live? Why do we die? Where do we come from? What is going to happen to us? What is love? Why do we make war? I sometimes say that these are questions that transcend perception; they concern issues that are too big or too small to be perceived by our five principal sensory systems.

Somewhat surprisingly, the existential intelligence does reasonably well in terms of our criteria. Certainly, there are individuals—philosophers, religious leaders, the most impressive statesmen—who come to mind as high-end embodiments of existential intelligence. Existential issues arise in every culture—in religion, philosophy, art, and the more mundane stories, gossip, and media presentations of everyday life. Indeed, in any society where questioning is tolerated, children raise these existential questions from an early age—though they do not always listen acutely to the answers! Moreover, the myths and fairytales that they gobble up speak to their fascination with existential questions.

My hesitation in declaring a full-blown existential intelligence comes from the dearth, so far, of evidence that parts of the brain are concerned particularly with these deep issues of existence. It could be that there are regions—for example, in the inferotemporal lobe—that are particularly crucial for dealing with the Big Questions. However, it is also possible that existential questions are just part of a broader philosophical mind—or that they are simply the more emotionally laden of the questions that individuals routinely pose. In the latter instances, my conservative nature dictates caution in giving the ninth place of honor to existential intelligence. I do mention this candidate intelligence in passing, but, in homage to a famous film directed by Federico Fellini, I shall continue for the time being to speak of "8 1/2 intelligences."

## THE UNIQUE CONTRIBUTIONS OF THE THEORY

My belief is that these multiple human faculties, the intelligences, are to a significant extent independent of one another. Research with brain-damaged adults repeatedly demonstrates that particular faculties can be lost while others are spared. This independence of intelligences implies that a particularly high level of ability in one intelligence, say mathematics, does not require a similarly high level in another intelligence, like language or music. This independence of intelligences contrasts sharply with traditional measures of IQ that find high correlations among test scores.

I speculate that the usual correlations among subtests of IQ tests come about because all of these tasks in fact measure the ability to respond rapidly to items of a logical-mathematical or linguistic sort; these correlations might be substantially reduced if one were to survey in a contextually appropriate way—what I call "intelligence-fair assessment"—the full range of human problem-solving skills.

Until now, I may appear to have suggested that adult roles depend largely on the flowering of a single intelligence. In fact, however, nearly every cultural role of any degree of sophistication requires a combination of intelligences. Thus, even an apparently straightforward role, like playing the violin, transcends a reliance on musical intelligence.

As an example: To become a successful violinist requires bodily-kinesthetic dexterity and the interpersonal skills of relating to an audience and, in a different way, choosing a manager; quite possibly it involves an intrapersonal intelligence as well. Dance requires skills in bodily-kinesthetic, musical, interpersonal, and spatial intelligences in varying degrees. Politics requires an interpersonal skill, a linguistic facility, and perhaps some logical aptitude.

Inasmuch as nearly every cultural role requires several intelligences, it becomes important to consider individuals as a collection of aptitudes rather than as having a singular problem-solving faculty that can be measured directly through pencil-and-paper tests. Even given a relatively small number of such intelligences, the diversity of human ability is created through the differences in these profiles. In fact, it may well be that the "total is greater than the sum of the parts." An individual may not be particularly gifted in any intelligence; and yet, because of a particular combination or blend of skills, he or she may be able to fill some niche uniquely well. Thus it is of paramount importance to assess the particular combination of skills that may earmark an individual for a certain vocational or avocational niche.

In brief, MI theory leads to three conclusions:

1. All of us have the full range of intelligences; that is what makes us human beings, cognitively speaking.
2. No two individuals—not even identical twins—have exactly the same intellectual profile. That is because, even when the genetic material is identical, individuals have different experiences, and those who are identical twins are often highly motivated to distinguish themselves from one another.
3. Having a strong intelligence does not mean that one necessarily acts intelligently. A person with high mathematical intelligence might use her abilities to carry out important experiments in physics or create powerful new geometric proofs, but she might waste these abilities in playing the lottery all day or multiplying 10-digit numbers in her head.

## CONCLUSION

I believe that in our society we suffer from three biases, which I have nicknamed "Westist," "Testist," and "Bestist." "Westist" involves putting certain Western cultural values, which date back to Socrates, on a pedestal.

"Bestist" is a not very veiled reference to a book by David Halberstam called *The Best and the Brightest*. Halberstam referred ironically to figures, such as Harvard faculty members, who were brought to Washington to help President John F. Kennedy and in the process launched the Vietnam War. Any belief that all the answers to a given problem lie in one certain approach, such as logical-mathematical thinking, can be very dangerous. Current views of intellect need to be leavened with other, more comprehensive points of view.

"Testist" denotes the obsession in contemporary society with tests—particularly the short-answer tests that can be administered quickly and scored even more quickly.

We are all different from one another in significant measure because we all have different combinations of intelligences. If we recognize this, I think we will have at least a better chance of dealing appropriately with the many problems that we face in the world. If we can mobilize the spectrum of human abilities, not only will people feel better about themselves and more competent; it is even possible that they will also feel more engaged and better able to join the rest of the world community in working for the broader good. Perhaps if we can mobilize the full range of human intelligences and ally them to an ethical sense, we can help to increase the likelihood of our survival on this planet, and perhaps even contribute to our thriving.

For more discussion of allying our multiple intelligences to an ethical sense, please see the papers on the Good Work Project and the Good Project in Essays 28 and 29.

## REFERENCES

Gardner, H. (1983). *Frames of mind: The theory of multiple intelligences*. Basic Books.
Gladwin, T. (1970). *East is a big bird: Navigation and logic on Puluwat atoll*. Harvard University Press.
Halberstam, D. (1972). *The best and the brightest*. Random House.
Keller, E. F. (1983). *A feeling for the organism*. W. H. Freeman and Company.
Lash, J. P. (1980). *Helen and teacher: The story of Helen Keller and Anne Sullivan Macy*. Delacorte Press.
Menuhin, Y., & Davis, C. W. (1979). *The music of man*. Methuen.
Ruth, B., & Considine, B. (1948). *The Babe Ruth story*. Pocket Books.
Woolf, V. (1985). *Moments of being: A collection of autobiographical writing of Virginia Woolf*. Mariner Books Classics.

# A "Smart" Lexicon

"John is so smart"—"John's a real dummy"

In English, and no doubt in other languages as well, individuals have been so characterized for many centuries. Indeed, for at least a century, psychology has provided a way to measure "smartness" or "dumbness." The IQ test provides a reasonable measure of how an individual is likely to do in school—particularly in a modern Western secular school. Of course, last year's grades or class standing provide equally helpful (or damaging) predictions.

A complementary language also emerged. In addition to being "smart" or "intelligent" or "quick-witted," individuals could be characterized as "talented." Perhaps John did not do well on an IQ test, but he might have been a talented artist, or humorist, or dancer or mechanic, or salesperson.

Critiques moved into second gear when psychologists—including me—proposed alternative ways of thinking about intellect (see Essay 17). Perhaps the best known is Daniel Goleman (1995), who introduced the idea of emotional intelligence.

The plurality of intelligence(s) is now widely acknowledged. We should say that "John is talented at school" or "John is not talented at this kind of school at this time." And within school, we should say that "Jane is good at math but not at language," or "Jane is good in history but not in physics" . . . or vice versa. I would go further—the distinction between "intelligence" and "talent" does not stand up to scrutiny. Either we should call all kinds of high abilities "talents" (including verbal and mathematical abilities), or we should call them all "intelligences." There is no principled distinction between a talent and an intelligence.

In what follows, with the hope that they will be useful, I introduce further distinctions in the lexicon of unusual human performance(s).

*Gifted/Prodigious/Talented:* In every form of intellect, some individuals will stand out from an early age. Some might be gifted in a particular area; others are so gifted in one or more areas—"off the charts"—that we call them prodigies.

*Expert:* In every known area of performance, there are adult standards for excellence. We call individuals who achieve those standards "experts." One characteristic of individuals who are described as gifted is that they will likely reach a level of expertise or mastery at an early age.

Full stop. Giftedness and expertise pertain to spheres of knowledge and performances that are widely recognized within a society. The following categories pertain to performances that fall outside that categorization.

*Creative/Creativity:* Some individuals and some performances move or even "hijack" domains of knowledge and practice in new directions. Such innovation can occur planfully or by accident. One can strive to be creative, but fail; one can engage in ordinary practice and find that one has in fact been creative.

In any event, as formulated persuasively by Mihaly Csikszentmihaly (1988), creativity does not primarily reflect intent; it reflects achievement. Only if a performance is deemed notable by the relevant communities, and only if the performance actually affects the subsequent standards of that community, does it merit the descriptor "creative."

Importantly, intelligence(s) and creativity(ies) are not the same. One can be highly intelligent in a domain but not creative; or one can be creative without being especially intelligent.

So far, what I've written draws on earlier work by conceptualizers of unusual performance. Recently I have become interested in a human capacity in which I personally seem to have expertise—the capacity to synthesize (see Essays 35–36).

A final entry in this lexicon:

*Polymath:* We use this term to describe someone who is knowledgeable in two or more areas: for example, a student who scores high in all academic areas; a scientist who is an expert in biology, mathematics, and music; or a painter who is also a composer and a poet. Polymaths impress us, but they are not the same as synthesizers. The polymath knows many things, but may show neither inclination to draw them together nor any aptitude to do so. The synthesizer surveys or investigates many areas, including ones in which (s)he has little training, with the motivation to tie the strands together in an illuminating way.

We might say that the synthesizer has a purpose in mind and needs a method to tie together lines of expertise; the polymath picks up knowledge easily and can display it readily but need not have either a purpose or method.

Of course, polymaths who are also synthesizers are a valuable resource. Whether they can be detected at early age, and whether those two forms of competence can be inculcated and intertwined over the course of development, is an important question—both for educators and for those concerned about the future of our planet. In both cases, of course, these talents need to be yoked to positive ends—in my terms, to the pursuit of good work and good citizenship.

Starting with the single construct of intelligence, psychologists and educators have proposed and provided evidence for various forms of cognition. In this essay, I have proposed an additional form—synthesizing ability—and sought to begin the process of integration among the various forms of intellect.

# REFERENCES

Csikszentmihaly, M. (1988). Society, culture, and person: A systems view of creativity. In R. J. Sternberg (Ed.), *The nature of creativity* (pp. 325–339). Cambridge University Press.

Goleman, D. (1995). *Emotional intelligence.* Bantam.

# Artistic Intelligences

*In the early 1980s, I was making a transition in my own work—from a focus on artistic and symbolic development and breakdown to an exploration of the theory of multiple intelligences. Naturally, I became interested in the relationship between these two lines of work. In this essay, I attempt to think through that relationship. The chapter suggests that any intelligence can be put to artistic uses, but that there is no necessary relation between artistry and intelligence. Musical intelligence underlies the work of Bach and the Beatles, but it can also be used to arouse one in the morning (trumpet reveille) or to soothe one seated in the dentist's chair.*

Scratch a school principal or, for that matter, an experimental psychologist, and you are likely to encounter a standard view of how humans participate in the arts. From dance to drama, the arts are seen as matters of emotion—arising from and stimulating feelings. Some individuals are endowed with talents in the arts, and, if so blessed, they simply wait until inspiration strikes. No realm of experience seems further away from formal schooling, from rationality, from scientific progress: how appropriate it is that the arts are listed in the newspaper as amusements and that one enters a concert hall or a museum with the same unreflective awe that one brings to church. Scratch an unreflective or defensive artist and you are likely to garner yet further testimony in support of such an affective and inspirational view of the arts.

But most artists know better. They must certainly be aware of the training and discipline that enters into and permeates their craft, of the difference between occasional inspiration and daily toil. Indeed, in their own informal conversations, artists speak frequently about the development and deployment of a wide range of abilities and skills. Yet there has been a virtual conspiracy of silence among artists concerning the arduous training and the keen mental efforts involved in artistic practice.

Given these commonly encountered opinions both in and outside "the trade," it is not surprising that the prevailing wisdom about participation in the arts was for many years studiously *non-* (or even *anti-*) cognitive. But in the past few decades, there has been significant change in the way in which aestheticians, artists, art educators, and others conceptualize the activities of artistic creators, performers, and audience members. Art is being seen anew—or once again—as a matter of the mind. Some credit should go to the zeitgeist, to the rise of the cognitive sciences, which sometimes threaten to overwhelm all human-oriented disciplines. Considerable credit should go to a few insightful commentators—Ernst Gombrich, Leonard Meyer, I. A. Richards, my venerated teachers Nelson Goodman and Rudolf Arnheim. These scholars have stressed (and exemplified) the cognitive facets of artistry. And part of the credit should go to that small but spirited flank of empirical researchers who have provided detailed evidence in support of such a cognitive view.

Just what is entailed in a cognitive view of the arts? As we have developed this perspective at Harvard Project Zero, artistry is first and foremost an activity of the mind. Like much other mental activity, artistic perception and production involves the use of symbols—a deployment that may well constitute the hallmark of human cognition. But artistry centers on the use of *certain kinds of symbols* (for example, paintings rather than chemical formulae), which are used in certain kinds of ways. To attain competence in the arts, it is necessary to gain literacy with these symbol systems.

And so the artistically competent individual is one who is able to "read" and to "write" symbols in such realms as literature, music, or sculpture. In this cognitive view, the role of the emotions is certainly acknowledged, but emotions are seen as aiding in the processes of symbol encoding and decoding, rather than as somehow opposed to "sheer" cognitive activity. That different works may have different value or merit is also acknowledged, but attention focuses on how one interacts with (or "comes to know") artistic symbols—not on whether one symbol is intrinsically better or worse, more or less beautiful than another.

In light of these considerations, a research agenda follows. The investigator of artistic knowledge studies the ways in which skilled (and unskilled) individuals handle artistic symbols: the goals they set, the problems they encounter, the steps through which they pass in fashioning or interpreting artistic symbols. Those interested in human development probe the stages through which children pass in gaining artistic mastery. Those interested in adult performance compare novices with experts, or study highly competent performance on a moment-by-moment basis. And those whose interests focus on educating artistic vision examine various methods for enhancing an individual's capacity to encode and decode artistic symbols.

Such is the program we have attempted to follow at Harvard Project Zero, a program which, happily, is being embraced by an increasing number of researchers (see Essay 3).

Having sketched something of the contour of artistic symbolization, its adult facets, its development trajectory, its educational regimen, we (as well as other researchers) have been thrown back to an important question, which, for a while, it was prudent to bracket. This is the question of the relationship between artistic and other forms of knowledge. How do the arts relate to other pursuits—to business, to athletics, to politics, and, above all, to the sciences? Are the arts a domain adrift, perhaps even a domain that has captured one half of the brain, or one half of the mind? Or does the cognitive division of labor prove more complex, with the relations between the arts and/or pursuits a complex and still largely uncharted territory?

To sort out these questions, I have recently reviewed a large body of the literature about human cognition that has accrued over the past decades. My special goal has been to cull insights from bodies of knowledge that had not hitherto been brought to bear upon one another. Included in this survey have been reviews of the development of cognition; the breakdown of cognitive capacities following various kinds of brain injury; the nature of abilities (and disabilities) in special populations, including prodigies, savants, autistic children, and others who exhibit unusual cognitive profiles; and the cross-cultural literature chronicling which cognitive abilities are valued in diverse societies, as well as relevant (though necessarily scattered) information about the evolution of cognition and about the intellectual capacities of other species. By means of this review of various lines of evidence, I have sought to determine whether there are certain cognitive proclivities that characterize human beings—and I have dubbed these the "multiple intelligences" (see Essay 17).

In affixing these relatively colorless labels of different kinds of intelligence, I am already classifying these potentials in a culturally tinted way. My claim, expressed more precisely, is that humans possess a set of semi-autonomous information-processing devices—one may think of them as "dumb" but reliable computers—which, when exposed to certain forms of environmental information, will carry out certain kinds of operations upon that information. The computational "core" of linguistic intelligence is phonological and syntactic analysis; the "core" of musical intelligence is rhythmic and pitch analysis; the "core" of logical mathematical intelligence is the perception of certain recurrent patterns, including numerical patterns; and so on.

Human beings live in cultures; these cultures can survive only if certain roles are filled and if certain functions are carried out. One means of survival is to ensure that these critical functions are passed on from one generation to the next. For this transmission to occur, various intellectual potentials must be mobilized. In my view, this mobilization occurs through the invention and dissemination of various kinds of symbolic products—books and speeches, pictures and diagrams, musical compositions, scientific theories, games, rituals, and the like.

For educational as well as scientific purposes, an analysis in terms of symbolic products turns out to be a judicious undertaking. On the one hand, these entities are sufficiently tangible so that the culture can assess whether roles are being fulfilled and knowledge transmitted. On the other hand, symbolic products are susceptible to brute information processing in the sense that "computational devices" can be brought to bear on them. To put it concretely, the human brain is equipped to process stories, while at the same time these stories prove an excellent cultural means of transmitting knowledge from one generation to another. The same can be said of number systems, musical songs, or religious rites. The "socialization" or "enculturation" of the aforementioned human intellectual proclivities is, to my mind, the major task of our education system.

In my view, the "intelligences" are not, either separately or jointly, preordained to be involved in the arts, the sciences, or any particular specific cultural area. Instead, they are raw computational mechanisms, which can be marshaled by artistic symbols or, equally, for other kinds of symbols, and other kinds of ends, when that seems indicated.

Thus, to take the case of language, there is no particular reason why an individual's linguistic potential needs to be harnessed in the service of metaphor, poetry, stories, or dramas. Yet, clearly, if these kinds of symbolic products are available, and if the culture (via its members, institutions, or values) chooses to highlight the literary use of language, this particular artistic faculty will be developed. Visual-spatial intelligence is similarly "blind"; it can be exploited for geometry, physics, engineering, or sailing, on the one hand, or it may be marshaled in the production of sculpture, painting, or choreography, on the other. Indeed, it turns out that each of the "multiple intelligences" I have proposed can be entrained in artistic activities: Some, like music, are typically involved in this way, whereas others, like logical-mathematical abilities, are only rarely so deployed. But there is no imperative that any given intelligence be sculpted into an aesthetic form; that turns out to be an accident of cultural history.

Such a portrayal of human intellectual competences has a number of implications for the way that we think of mind. For a start, this portrait proves far more pluralistic than the cultural stereotype of a single form of intelligence with which each individual is endowed (for better or worse) at birth. Presumably, individuals do differ in their potential in each of these domains, but there is ample room for developing several intelligences, alone or in combination; and an individual's peculiar profile of development will give rise to a wide array of complex skills.

This framework also highlights the extent to which the surrounding culture determines the uses to which one's raw intellectual potentials are put. Clearly a society can opt for a highly artistic diet—as Bali or Japan appear to have done—or a society can choose to adopt a far more scientific, technological, or economic regimen, thereby minimizing the incidence and importance of artistic symbolic products.

Educational implications follow as well. One interesting implication is that we should stop regarding perception, memory, and learning as extremely general capacities, applicable equivalently to every manner of content. According to my analysis, *there may be specific forms of perception, memory, and the acquisition of new knowledge for each of the intellectual competences.* At the very least there is no reason, *a priori,* to assume that a heightened memory in one particular domain implies anything about one's mnemonic capacities in a neighboring (or contrasting) one.

Another implication pertains to early detection of an individual's intellectual profile. My own guess is that individuals differ in their potentials in various domains and that an individual's strengths can be identified quite early: ability to recognize patterns, and then to retain them, is probably discernible in the first years of life, and these capacities may serve as a sensitive measure of one's inherent talents in one or another intellectual domain. Such an assessment of an individual's "intellectual profile" should be thought of as informative rather than limiting. (See Essay 11 on crystallizing experiences in *The Essential Howard Gardner on Education*.)

Armed with this kind of knowledge, parents and educators have the option of developing a child's strengths, and of supplementing weaker areas, either through special training or through the use of prosthetics, which can often supplant modest endowment in a given intellectual domain. Or they can seek to do both!

According to my analysis, we have tended in our own society to accord excessive weight to linguistic and logical-mathematical intelligences, while giving relatively short shrift to other intellectual domains. Our aptitude and achievement tests are also far more sensitive to accomplishments in the aforementioned domains.

In order to test an individual's ability in the bodily or musical domains, for example, it is clearly inadequate simply to use paper-and-pencil measures that can be administered in an hour . . . or less!

Rather, we should ask individuals to participate in activities that actually call in significant measures on these "intelligences"—to dance, to play a sport, to sing, to learn an instrument. It should be possible to develop intrinsically compelling activities (for example, simple games) that allow a ready assessment of an individual's interest, and potential for development in a given intellectual domain. Identifying such "crystallizing experiences" in the arts is an important task for educators. Properly deployed, such experiences can be used to assess a child's "zone for potential development" and to increase the likelihood that the assessed potential gets actualized.

This novel view of intelligence suggests that the arts may be especially suited to encompass the range of individual intellectual profiles. No matter how idiosyncratic an individual's intellectual skills, there should be art forms and products that can mobilize them. The menu of choices—literature and music, painting and dancing, acting, carving and sculpture—prove sufficiently variegated to allow virtually every individual to gain pleasure and to achieve competence. To be sure, art educators should not force-feed these activities; that would be as misguided as a diet of *all* sciences, *all* sports, or *all* commercial endeavors. But to the extent that we wish children to have the opportunity to develop their full range of intellectual potentials, it is virtually an imperative that we facilitate involvement in one or more art forms.

## REFERENCE

Gardner, H. (1983). Artistic intelligences. *Art Education (Reston)*, 36(2), 47–49.

# Who Owns Intelligence?

While the theory of multiple intelligences was becoming widely known (and appreciated) in education, and widely known (and critiqued) in academic psychology, it was not yet known by the wider public. I welcomed the invitation from *The Atlantic Monthly* (now called *The Atlantic*) to introduce the background and central claims of the theory—including the addition of the natural intelligence. This platform also gave me the opportunity to place MI theory in a broader context, to address various critiques, to comment on emotional intelligence (a concept that was far better known), and to relate intelligence to other areas of study, like emotions, morality, and creativity.

Here, without repeating the "basics" of MI theory (Essays 17–19 in this volume) and *The Essential Howard Gardner on Education* (Essays 8–15), I reprint excerpts that seemed to resonate with a broader public.

As a psychologist, I've always assumed that my fellow psychologists and I owned the concept of intelligence. Of course, I realize that the word "intelligence" is used by those in the diplomatic community, and that it has a proper place in our idle chatter. But when it comes to the scientific study of intelligence, psychologists have had a virtual monopoly on the territory . . . at least until now.

Alfred Binet, the inventor of the IQ test, is a hero to many psychologists. He was a keen observer, a careful scholar, and an inventive technologist. At the start of the 20th century he devised the instrument that is often considered psychology's greatest success story. Millions of people who have never heard Binet's name have had their fates determined by instrumentation that the French psychologist inspired. And thousands of measurement specialists—called psychometricians—earn their livings courtesy of Binet's invention.

But while successful over the long run, the psychologist's version of intelligence is now facing its biggest threat. Many scholars and observers—and even some iconoclastic psychologists—feel that intelligence is too important to leave to the psychometricians. Experts are extending the breadth of intelligence—proposing many intelligences, including emotional intelligence and moral intelligence. They are experimenting with new methods of ascertaining intelligence, including ones that avoid tests altogether in favor of direct measures of brain activity. They are forcing society to confront a number of questions: What is intelligence? How ought it to be assessed? And how do our notions of intelligence fit with what we value about human beings? In short, as my title suggests, experts are competing for the "ownership" of intelligence in the next century.

*    *    *

The outline of the psychometricians' success story is well known. Binet's colleagues in England and Germany contributed to the conceptualization and instrumentation of intelligence testing, which soon became known as the IQ test. (An Intelligence Quotient designates the ratio between mental age and chronological age. Clearly it is preferable for a child to have an IQ of 120—where one is putatively smarter than one is old—than an IQ of 80—where one is older than one is smart.) And, like other Parisian fashions of the period, the intelligence test migrated easily to the United States. First used to determine who was feebleminded, it was soon used to assess "normal children," to identify the "gifted," and to determine who was fit to serve in the Army. By the 1920s, the intelligence test had become a fixture in educational practice in the United States and through much of Western Europe.

Early intelligence tests were not without their critics. Many enduring concerns were first raised by the influential journalist Walter Lippmann in a series of published debates with Stanford University's Lewis Terman, the "father" of IQ testing in America. Lippmann noted the superficiality of the questions, their possible cultural biases, and the "high-stakes" risks of assessing an individual's intellectual potential via a brief oral or paper-and-pencil measure. IQ tests were also the subject of many jokes and cartoons. Still, by sticking to their trade, the psychometricians were able to defend their instruments, even as they made their way back and forth among the halls of academe, their testing cubicles in schools and hospitals, and the vaults in their banks.

Perhaps surprisingly, the conceptualization of intelligence did not advance much in the decades following Binet and Terman's pioneering contributions. Intelligence testing came to be seen, rightly or wrongly, as primarily a technology for selecting individuals to fill academic or vocational niches. In one of the most famous—if irritating—quips about intelligence testing, the influential Harvard psychologist E. G. Boring declared, "Intelligence is what the tests test." As long as these tests did what they were supposed to do—that is, give some indication of school success—it did not seem necessary or prudent to probe too deeply into their meanings or to explore alternative views of the matter.

Psychologists of intelligence have argued chiefly about three questions:

The first: *Is intelligence singular, or are there various more or less independent intellectual faculties?* Hedgehog purists[1]—ranging from the English psychologist Charles Spearman at the turn of the 20th century to his latter-day disciples

---

1. The humanist scholar, Isaiah Berlin, was fond of contrasting foxes who knew many little things with hedgehogs who knew one big thing. The distinction was initially proposed by the classical Greek poet Archilochus.

Richard Herrnstein and Charles Murray (of *The Bell Curve* [1994] fame)—defend the notion of a single supervening "g" or general intelligence.

Foxlike pluralists—ranging from Chicago's L. L. Thurstone, who posited seven vectors of the mind, to California's J. P. Guilford, who discerned 150 factors of the intellect—construe intelligence as composed of some or even many dissociable components. In his much-cited *The Mismeasure of Man* (1981), paleontologist Stephen Jay Gould argued that the conflicting conclusions reached on this issue simply reflect alternative assumptions about statistical procedures rather than "the way the mind is." Still, psychologists continue to debate this issue, with a majority sympathetic to a single, "general intelligence" perspective.

The lay public is more interested in a second contentious question: *Is intelligence (or are intelligences) largely inherited?* It should be noted that this is by and large a Western question. In the Confucian societies of East Asia, it is assumed that individual differences in endowment are modest and that differences in achievement are due largely to effort. In the West, however, there is much sympathy for the view that intelligence is inborn and that there is little one can do to alter one's quantitative intellectual birthright.

Studies of identical twins reared apart provide surprisingly strong support for the "heritability" of psychometric intelligence. That is, if one wants to predict someone's score on an intelligence test, it is more relevant to know the identity of the biological parents (even if the child has not had appreciable contact with them) than the identity of the adoptive parents. By the same token, the IQs of identical twins are more similar than the IQs of fraternal twins. And, contrary to common sense (and political correctness), IQs of biologically related individuals grow more similar in the later years of life, rather than more different. Still, because of the intricacies of the discipline of behavioral genetics, and the difficulties of conducting valid experiments with human childrearing, one still finds those who defend the proposition that intelligence is largely environmental, as well as those who believe that researchers cannot answer this question at all.

One other question has intrigued lay individuals and psychologists: Are intelligence tests biased? If one looks at early intelligence tests, the whopping cultural assumptions built into certain items are evident. There are obvious class biases—who except the wealthy should be able to answer a question about polo? There are also subtler nuances—while ordinarily it makes sense to turn over money found on the street to the police, what happens in the case of a hungry child? Or with respect to a police force that is known to be hostile to members of one's own minority group? Only the canonical response to such a question would be scored as correct.

Psychometricians have striven to remove the obviously biased items from such measures. Yet it is far more difficult to deal with biases that are built into the test situation itself. For example, an individual's background certainly figures into his reactions to being placed in an unfamiliar locus, instructed by an interrogator dressed in a certain way, and having a printed test booklet thrust into his hands. The biases prove even more acute in cases where an individual knows that her intellect is being measured, and where she belongs to a racial or ethnic group that is widely considered to be less smart than the dominant social group.

Talk of bias touches on the frequently held assumption that tests in general, and intelligence tests in particular, are inherently conservative instruments— tools of the establishment. It is therefore worth noting that many test pioneers thought of themselves as progressives in the social sphere. They were devising instruments that could reveal individuals of talent, even if those persons came from "remote and apparently inferior backgrounds" (expression from a handbook of college admissions of the period). And occasionally the tests did discover intellectual "diamonds in the rough." More often, however, the tests picked out individuals of privilege—the correlation between zip code and IQ (and SAT score) is high. The still-unresolved question of the casual relation between IQ and social privilege has stimulated many a dissertation across the social sciences.

Paradoxically, one of the clearest signs of the success of intelligence tests is that they themselves are not widely administered anymore.

However, despite this apparent setback, intelligence testing, and the line of thinking that underlies it, have actually won the war. Many widely used scholastic measures—chief among them the SAT—are thinly disguised intelligence tests—almost clones thereof—that correlate highly with scores on standard psychometric instruments. Virtually no one raised in the developed world today has remained untouched by Binet's deceptively simple invention of a century ago.

*   *   *

Secure in practice, the concept of intelligence has in recent years undergone its most robust challenge since the days of Walter Lippmann. For the first time in many years, the Intelligence Establishment is clearly on the defensive. It seems likely—particularly at a time of increasingly intelligent computer systems—that we will develop different ways of thinking about intelligence.

One evident factor in the rethinking of intelligence is the perspective introduced by those scholars who are not psychologists. Anthropologists have noted the parochialism of the Western view of intelligence. Some cultures do not even have a concept called intelligence, and others define intelligence in terms of traits that we in the West might consider odd—obedience, or good listening skills, or moral fiber, for example. Neuroscientists are skeptical that a single or unitary form of intelligence is consistent with the highly differentiated and modularized structure of the brain. Computer scientists have devised programs deemed to be intelligent; these programs often go about problem-solving in ways quite different from those followed by human beings or other animals.

Even within the field of psychology, the natives have been getting restless. Unquestionably the most restless is Yale psychologist Robert Sternberg. A prodigious scholar, Sternberg has written dozens of books and hundreds of articles, the majority of them focusing on intelligence in one or another way. Sternberg began with the strategic goal of understanding the actual mental processes mobilized on standard test items, such as the solving of analogies. But he soon went beyond the components of standard intelligence testing by insisting on two hitherto neglected forms of intelligence: the "practical" ability to adapt to varying contexts (as we all must in these days of divorcing and downsizing); and the capacity to automate familiar activities so that one can deal effectively with novelty and display "creative" intelligence.

My own work on "multiple intelligences" also eschews the psychologists' credo of operationalization and test-making. Instead, I began by asking two questions: The Evolutionary Question, "How did the human mind/brain evolve over millions of years?"; and the Comparative Question, "How can we account for the diverse skills and capacities that are or have been valued in different communities around the world?"

Armed with those questions, and a set of eight criteria for what "counts" as an intelligence, I have concluded that all human beings possess at least eight intelligences: linguistic and logical-mathematical (the two most prized in school and the ones central to success on intelligence test-type instruments); musical; spatial; bodily-kinesthetic; naturalist; and two forms focused on human beings, interpersonal and intrapersonal (see Essays 17–19).

I make two complementary claims about intelligence. The first is a universal claim. We all possess these eight intelligences—and possibly more. Indeed, rather than seeing the human as a "rational animal," I offer a new definition of what it means to be a human being, cognitively speaking: *homo sapiens sapiens* is the animal that possesses these eight forms of mental representation.

My second claim concerns individual differences. Due to the accidents of heredity, environment, and their interactions, no two of us exhibit the same intelligences in precisely the same proportion and blend. Our "profiles of intelligence" differ from one another. This fact poses intriguing challenges and opportunities for our educational system. We can either ignore these differences and pretend that we are all the same; historically speaking, that is what most educational systems have done. Or we can fashion an educational system that tries to honor and nurture these differences, individualizing instruction and assessment as much as possible.

\*    \*    \*

Surveying the landscape of intelligence, I discern three sets of struggles between opposing forces. To put my cards on the table, I feel that the three struggles are interrelated; the first struggle provides the key to the others; and there is an optimal way in which to resolve the ensemble of struggles.

The first struggle concerns the *breadth* of our definition of intelligence. One camp consists of *traditionalists* who believe in a single form of intelligence, one that basically predicts success in school and in school-like activities. Arrayed against the traditionalists are the *progressive pluralists*. These individuals believe that there are many forms of intelligence. And some of these pluralists would like to broaden the definition of intelligence considerably, to include the abilities to create, to lead, and/or to stand out in terms of emotional sensitivity or moral excellence.

The second struggle concerns the *assessment of intelligence*. Again, one readily encounters a traditional position. Once sympathetic to paper-and-pencil tests, the traditionally oriented practitioner now looks to computers to provide the same information more quickly and more accurately.

But other positions abound. "Purists" disdain psychological tasks of any complexity, preferring to look instead at reaction time, brain waves, and other "purer" physiological measures of intellect. "Simulators" move in the opposite direction, to more realistic life-sized measures that closely resemble the actual abilities that are prized. And "skeptics" warn against the continuing expansion of testing. They emphasize the damage often done to individual life chances and self-esteem by a regimen of high-stake psychological testing. And they call instead for less technocratic, more humane methods. These range from self-assessment, to the examination of portfolios of student work, to selection in the service of social equity.

The final struggle concerns the *relationship between intelligence and what qualities we value in human beings*. While no one would baldly equate intellect and human value, nuanced positions have emerged on this issue. Some see intelligence as closely related to a person's ethical and value system; they anticipate that brighter individuals are more likely to appreciate moral complexity and to behave judiciously. Some call for a sharp distinction between the realm of intellect, on the one hand, and character, morality, or ethics, on the other. Society's ambivalence on this issue can be discerned in the figures that become heroes in the media. For every Albert Einstein or Bobby Fischer who is celebrated for his intellect, there is a fictional Forrest Gump or Chauncey Gardiner, who is celebrated precisely for those human (and humane) traits that would never be captured on any kind of an intelligence test.

Reflecting on these struggles, I believe that the most pivotal battlefield is likely to be the one in which the new dimensions and boundaries of intelligence are thrashed out. Thanks to the work of the past few decades, the stranglehold of the psychometricians has at last been broken. This is a beneficent development. Yet now that the Scylla of the psychometricians has been overcome, we risk succumbing to the Charybdis of "anything goes"—emotions, morality, creativity all become absorbed into the "New Intelligence." The challenge is to chart a concept of intelligence that reflects new insights and discoveries and yet can withstand rigorous scrutiny.

An analogy may help. One can think of the scope of intelligence as represented by an elastic band. For many years, the definition of intelligence went unchallenged and the band seemed to have lost its elasticity. Some of the new definitions expand the band so that it has become quite taut and resilient; and yet earlier work on intelligence is still germane. Other definitions so expand the band that it finally snaps—and the earlier work on intelligence can no longer be drawn upon.

Until now, the term *intelligence* has been limited largely to certain kinds of problem-solving involving language and logic—the kinds of skills at a premium in the lawyer or law professor. However, humans are able to deal with numerous other contents besides words, numbers, and logical relations—for example, the contents of space, music, the psyches of other human beings. Like the elastic band, conceptions of intelligence need to be expanded to include human skill in dealing with these diverse kinds of contents. And we must not restrict attention to the solving of problems that have been posed by others; we must consider equally the capacities of individuals to *fashion products* (like works of art, scientific experiments, or effective organizations) that draw on one or more of several human intelligences. The elastic band can accommodate such broadening as well.

As long as intelligences are restricted to the processing of "contents in the world," we avoid epistemological problems. So it should be. "Intelligence" should not be expanded to include personality, motivation, will, attention, character, creativity, and other important and significant human capacities. Such stretching is likely to snap the band of intelligence altogether.

Let's see what happens when one crosses one of these lines—for example, when one attempts to conflate intelligence with creativity. Beginning with a definition, we extend the descriptor "creative" to those individuals (or works or institutions) who meet two criteria: (1) they are innovative, and (2) their novelty is eventually accepted by a relevant community or domain.

No one denies that creativity is important—and, indeed, it may prove even more important in the future, when nearly all standard (algorithmic) procedures are carried out by computers. (See Essays 23 and 24). Yet creativity should not be equated with intelligence. An expert may be intelligent in one or more domain, but there is no necessity that he or she be inclined toward, or successful in, innovation. Similarly, while the ability to innovate clearly requires a certain degree of intelligence (or intelligences), there is not otherwise a significant correlation between measures of intellect and creativity. Indeed, creativity seems more dependent on a certain kind of temperament and personality (risk-taking, tough-skinned, persevering, and above all, having a lust to alter the status quo and leave a mark on society) than on efficiency in processing various kinds of informational content. By collapsing these categories together, we risk missing dimensions that are important but separate; and we may think that we are training (or selecting) one, when we are actually training (or selecting) the other.

Consider, next, what happens when one stretches intelligence to include good or evil attitudes and behaviors. By this incursion into morality, we are now confronting human values within a culture. There may be a few values that can be expressed generically enough so that they command universal respect: the Golden Rule (biblical-style) is one promising candidate! Almost every other value, however, turns out to be specific to cultures or subcultures—even such seemingly unproblematic ones as the unacceptability of incest, killing, or lying. Once one conflates morality and intelligence, one needs to deal with the widely divergent views of what is good and bad and why. Moreover, one must deal with the fact that individuals who score high on tests of moral reasoning often act immorally outside the test situation, even as courageous and self-sacrificing individuals turn out to be unremarkable on formal tests of moral reasoning and on intelligence tests. It is far preferable to construe intelligence itself as morally neutral, and then apply a different set of calipers to decide whether a given use of intelligence qualifies as moral, immoral, or amoral in a given context.

As I see it, no intelligence is moral or immoral in itself. One can be gifted in language and use that gift (as did Johann Wolfgang von Goethe) to write great verse, or (as did Josef Goebbels) to foment hatred. Mother Teresa and Lyndon Johnson, Niccolò Machiavelli and Mohandas Gandhi may have had equivalent degrees of interpersonal intelligence, but the uses to which they put their skills could not have been more varied.

One might respond by saying, "Perhaps there is an intelligence that determines whether or not a situation harbors moral considerations or consequences." I have less problem with such a formulation. Note, however, that the term "moral intelligence" loses much of its force. After all, Adolf Hitler or Joseph Stalin may well have had an exquisite sense of which situations were considered moral; however, either they did not care, or they embraced their own peculiar code of what counted as moral ("Eliminating Jews is the moral thing to do in quest of a pure Aryan society"; "Wiping out a generation is a necessary move if you want to establish a communist state").

*     *     *

I call, then, for a delineation of intelligence that includes the full range of contents to which human beings are sensitive; but one that, at the same time, designates, as off limits, such valued but separate human traits as creativity, morality, or emotional appropriateness. I believe that such a delineation makes scientific and epistemological sense; it reinvigorates the elastic band without stretching it to the breaking point; and it helps to resolve the two remaining struggles: how to assess, and what kinds of human beings to admire.

Once we decide to restrict intelligence to human information-processing and product-making capacities, we can make use of the established technology of assessment. That is, we can continue to use paper-and-pencil or computer-adapted testing techniques while looking at a broader range of capacities, such as musical sensitivity or the understanding of other persons. And we can avoid ticklish and possibly unresolvable questions about the assessment of values and morality that may well be restricted to a particular culture, and that may well change over time.

I turn to a final concern: the relationship between intelligence and what I will, for short, call virtue: those qualities of the human being that we admire and that we wish to hold up as examples to our own children and to other persons' children. No doubt, the desire to expand intelligence to encompass ethics and character represents a direct response to the general feeling that our society is lacking in these dimensions; the expansionist view of intelligence reflects the hope that if we transmit the technology of intelligence to these virtues, we might in the end secure a more virtuous population (see Essays 28 and 29).

I have already indicated my strong reservations about hijacking the word "intelligence" so that it becomes all things to all people—the psychometric equivalent of the true, the beautiful, and the good. Yet the problem remains: How, in a post-Aristotelian, post-Confucian era, one in which psychometrics looms large, do we think about the virtuous human being?

My analysis suggests one promising approach. We should recognize that intelligences, creativity, and morality—to mention just three desiderata—are separate. Each may require its own form of measurement or assessment, and some will prove far easier to assess in an objective manner than others. Indeed, with respect to creativity and morality, we are more likely to rely on overall judgments by experts than on any putative test battery. At the same time, there is nothing to prevent us from looking for individuals who combine more than one of these attributes: individuals who have musical and interpersonal intelligence; individuals who are psychometrically intelligent and creative in the arts; individuals who combine emotional sensitivity and a high standard of moral conduct.

In one of the most important (and oft-quoted) phrases of the 20th century, the British novelist E. M. Forster counseled us, "Only connect" (1910). I believe that, though well motivated, some expansionists in the territory of intelligence have prematurely asserted connections that do not exist. But I also believe that it is within our power as human beings to help forge connections that may be important for our physical and psychic survival.

Just how the precise borders of intelligence are drawn is a question that we can leave to scholars. But the imperative to broaden our definition of intelligence in a responsible way goes well beyond the academy. Who "owns" intelligence promises to be an even more critical issue in the next century than it has been in this era of the IQ test.

## REFERENCES

Forster, E. M. (1910) *Howards end.* G. P. Putnam's Sons.

Gould, S. J. (1981). *The mismeasure of man.* W. W. Norton & Company.

Herrnstein, R. J., & Murray, C. A. (1994). *The bell curve: Intelligence and class structure in American life.* Free Press.

# COGNITION

After completing the research and introducing and explaining multiple intelligences, I moved in new directions.

Given my training and longtime interest in interdisciplinary work, I decided to learn more about a newly emergent field called "cognitive science." After all, my major area of interest was "cognition," and I was intrigued that it had now given rise to a new field of study—one that sought to integrate concepts and findings from psychology, anthropology, linguistics, neurology, computer science, and—the grandfather of such studies—philosophy.

Cognitive science received impetus from two sources: (1) the combination of fields drawn on, at research centers around the country (and in the United Kingdom), to help win World War II; and (2) the decision of the Sloan Foundation, a science-focused philanthropy, to support new initiatives in science, such as neuroscience and behavioral economics.

And so, aided by a grant from Sloan, I went around the country and to Europe, spoke to leaders of these respective fields, and put together a book, *The Mind's New Science: A History of the Cognitive Revolution*.

### REFERENCE

Gardner, H. (1985). *The mind's new science: A history of the cognitive revolution*. Basic Books, Inc.

# Definition and Scope of Cognitive Science

In the course of proposing and founding a new field of knowledge, many individuals will formulate their own definitions. Indeed, since the term *cognitive science* first began to be bandied about in the early 1970s, dozens of scientists have attempted to define the nature and scope of the field. It therefore becomes important for me at the outset to state what I take cognitive science to be.

I define cognitive science as a contemporary, empirically based effort to answer longstanding epistemological questions—particularly those concerned with the nature of knowledge, its components, its sources, its development, and its deployment. Though the term cognitive science is sometimes extended to include all forms of knowledge, animate as well as inanimate, human as well as nonhuman, I apply the term chiefly to efforts to explain human knowledge. I am interested in whether questions that intrigued our philosophical ancestors can be decisively answered, instructively reformulated, or permanently scuttled. Today cognitive science holds the key to whether they can be.

Of the various features or aspects generally associated with cognitive scientific efforts, I consider five to be of paramount importance. Not every cognitive scientist embraces every feature, of course; but these features can be considered symptomatic of the cognitive-scientific enterprise. When all or most are present, one can assume that one is dealing with cognitive science; when few, if any, are present, one has fallen outside my definition of cognitive science. These features are dissected throughout my book *The Mind's New Science* (1985), but it is important to make an initial acquaintance with them at this point.

First of all, there is the belief that in talking about human cognitive activities, it is necessary to speak about *mental representations* and to posit a level of analysis wholly separate from the biological or neurological, on the one hand, and the sociological or cultural, on the other.

Second, there is the faith that *central to any understanding of the human mind is the computer.* Not only are computers indispensable for carrying out studies of various sorts, but more crucially, the computer also serves as the most viable model of how the human mind functions.

While the first two features incorporate the central beliefs of current cognitive science, the latter three concern methodological or strategic characteristics.

The third feature of cognitive science is the *deliberate decision to deemphasize certain factors that may be important for cognitive functioning but whose inclusion at this point would unnecessarily complicate the cognitive-scientific enterprise.* These factors include the influence of affective factors or emotions, the contribution of historical and cultural factors, and the role of the background context in which particular actions or thoughts occur.

As a fourth feature, cognitive scientists harbor the faith that *much is to be gained from interdisciplinary studies.* At present, most cognitive scientists are drawn from the ranks of specific disciplines—in particular, philosophy, psychology, artificial intelligence, linguistics, anthropology, and neuroscience (I refer to these disciplines severally as the "cognitive sciences"). The hope is that someday the boundaries between these disciplines may become attenuated or perhaps disappear altogether, yielding a single, unified cognitive science.

A fifth and somewhat more controversial feature is the claim that *a key ingredient in contemporary cognitive science is the agenda of issues, and the set of concerns, that have long exercised epistemologists in the Western philosophical tradition.* To my mind, it is virtually unthinkable that cognitive science would exist, let alone assume its current form, had there not been a philosophical tradition dating back to the time of the Greeks, especially Socrates and Aristotle.

# Scientific Psychology: Should We Bury It or Praise It?

*While my professional identity has always been primarily as a psychologist, I do not think of myself particularly as belonging to that trade: I am better described as a qualitative social scientist or, more recently, as a synthesizer about the mind.*

*But when I received the William James Award of the American Psychological Association, I used this honor as an opportunity to step back and to reflect on what I thought of psychology. I used the phrase "scientific psychology," though at the time all psychologists still fell under the tent of the APA.*

*As of 2024, those who see themselves as scientists belong to the Association for Psychological Science, while those who are primarily clinicians still congregate with the American Psychological Association. Academic politics!*

*In this essay, I contend that psychology is not an integrated science in the sense that biology, chemistry, or physics are sciences. And I think psychologists make a mistake in labeling themselves as scientists—as philosopher John Searle has famously quipped, "Anything that calls itself a science isn't a science."*

*Yet I respect the field of psychology and those who call themselves psychologists, whether or not they add the descriptor "science." And in this essay, I attempt to indicate what I see as the core of psychology—scientific or not—and what I see as likely to be absorbed by more traditional branches of science.*

Well over a century ago, William James signed a contract to write the first American textbook on psychology. As he wrote to his friend Thomas W. Ward, "I have blocked out some reading in physiology and psychology. It seems to me that perhaps the time has come for psychology to begin to be a science" (quoted in Feinstein, 1984, p. 313). As is well known, a task slated to be completed in 2 years dragged on for a dozen, but in the end, James had expounded on the subject in a way which has seldom if ever been equaled. Certainly, it is difficult to think of any other textbook that is read not only for pleasure but also for profit more than a century after its initial publication.

Tens of thousands of works in psychology have since been published, and psychology as a discipline—academic and practical—has achieved remarkable success. Yet, it is still not clear to many observers that the promise implied by a two-volume text in a new field called psychology has actually come to fruition.

Clearly advances have been made in many, if not most, of the topics treated by James and his immediate successors. But have these advances added up to a unified discipline whose components interrelate with one another? Are they worthy to be called a science in the same sense that biology, chemistry, and physics—or, for that matter, economics or demography—merit that label? Are there serious attempts to tie together the "micro" and "macro" levels, as are currently under way in the biological and the physical sciences?

In treating the possibility of psychology as a unified science, I am discussing a topic that William James would have found of interest. He himself had often voiced misgivings about the "confused and imperfect state" and the "antiscientific condition" of psychology. In my view, James's concerns have proven all too justified. Psychology has not added up to an integrated science, and it is unlikely ever to achieve that status. It no longer makes sense to discuss scientific psychology as a tenable long-term goal. What does make sense is to recognize important insights that have been achieved by psychologists; to identify the contributions that contemporary psychology can make to disciplines that may someday achieve a firmer scientific status; and, finally, to determine whether at least parts of psychology might survive as participants in a conversation that obtains across major disciplines.

Despite Immanuel Kant's dismal epitaph at the end of the 18th century, that a science of psychology was impossible, psychology proceeded to conquer much of the academic world.

What followed might be called, in the argot of Chinese dynasties, the Period of the Warring Schools: We had functionalism, behaviorism, structuralism, Gestalt psychology, learning theory, psychoanalysis, and a pack of other "isms"; we hosted the movements surrounding charismatic scientists like James J. Gibson, Clark Hull, Jean Piaget, and B. F. Skinner; and we experienced a number of worldly successes, such as the intelligence test, various indices of psychopathology, and the integrated realm that spans persuasion, advertising, and marketing. Psychology has become established as a potent societal force, with its departments, journals, institutions, and huge organizations, most prominent among them the 157,000-member American Psychological Association.

At least on the level of lip service, the dream of a unified psychology continues. It appears at the beginning and end of textbooks, though much less frequently in the intervening chapters. It surfaces as well in university catalogues and in the boilerplate statements of granting agencies. And occasionally, a scholar—more often an outsider or "independent researcher" than a practitioner of "normal science"—actually proposes a formula or "central dogma" for the field, one that purports to link all subfields and to bind the "micro" with the "macro." But for the most part, psychologists (like other academics) go about their daily research and writing without agonizing about the actual or potential coherence of their field.

Inasmuch as psychology gives little sign of cohering, we are faced with the following options.

1. We can simply close our minds to the possibility of disciplinary extinction and continue what we have been doing. No superbody is likely to denounce psychology as a fraud, and so we can maintain the status quo.
2. We can simply declare that psychology is a success—as it has been, according to many criteria—and swallow any lingering doubts that we might entertain.
3. We can hope that we are simply passing through a temporary phase of fragmentation and that some enterprising researcher, or some brilliant theorist, will discover the "golden thread" that will tie together the strands of our field.
4. We can claim that there has been an unjustified romanticization of other disciplines. After all, there are any number of subfields of biology; the geneticists or molecular biologists inhabit quite different worlds from the evolutionists, taxonomists, or paleontologists. And economics is at least as top-heavy as psychology with schools that struggle against one another.

There are certainly other options, but I favor a fifth. Let us recognize that fields of science evolve, often in unsuspected and unexpected ways. Nearly every field begins as philosophy; and psychology continues to foreground its philosophical origins more faithfully than any other discipline. There was a period 200 years ago—in the era of Immanuel Kant—when psychology seemed impossible; a set of discoveries in the 19th century that established a number of enduring psychological paradigms and concepts; and a complex of social and historical factors in the 20th century that earned psychology a place in virtually every academic environment.

## THE EMERGING DISCIPLINARY TOPOGRAPHY

Roughly paralleling breakthroughs in physics in the early decades of the 20th century, and the parallel advances in molecular biology at midcentury, the years at the close of our century can be well described as the coming-of-age of brain or neuroscience (see Essays 13–16).

If neuroscience will absorb much from the "lower regions" of psychology, an analogous kind of raid will be made by cognitive science—perhaps from the "top," perhaps more laterally (see Essay 21). This recently emerging science is a self-styled interdisciplinary field that, like traditional psychology, seeks to uncover the basic processes of thought; however, adopting the current vogue, cognitive scientists regard the computer as the most suitable model for all forms of cognition.

Neuroscience and cognitive science stand as the two behemoths threatening to absorb many settlements of science, including the mainstream of research in psychology. A recent hybrid, *cognitive neuroscience*, has even greater imperialistic design. It is certainly the case that other fields of psychology, such as social psychology, developmental psychology, or clinical psychology, are less at risk of immediate absorption; possibly because they lack easily transportable research paradigms, they can continue to evolve with less threat of a takeover by an interdisciplinary "corporate raider."

## THE SURVIVING CENTER

It may seem that, in this "Cook's tour" of the disciplinary topography of the future, we have drifted far away from William James and his view of psychology. But that is only because I have yet to mention those subjects—and those chapters—that were central in William James's own account (James, 1893).

I refer here to *Consciousness,* treated in Chapter 11; *The Self*—treated in Chapter 12; *Will* treated in Chapter 26, the concluding substantive chapter; and *Personality*, which, while rarely mentioned explicitly by James, is in fact an important presence in these chapters.

For James, the issue of the self or ego—its experiences, its internal and social aspects, its aspirations, and its evolution through life—is key in psychology. Indeed, when he heard that the psychoanalyst Sigmund Freud was coming to America, the ailing James made his way from Cambridge to Clark University in Worcester and declared to the Viennese visitor, "The future of psychology belongs to your work." As the historian H. Stuart Hughes comments, "there is no more dramatic moment in the intellectual history of our time" (1961, p. 113).

Since the time of James and Freud, the study of personality, self, will, and consciousness (hereafter, the "person-centered quartet") has occupied a paradoxical position within psychology. On the one hand, these topics are clearly central in any delineation of the field, and they occupy predictably pivotal spots in textbooks. And yet I must acknowledge that there is a slight embarrassment about these topics. While work continues on each of them, progress here is even less compelling than in other strands of psychology. Indeed, with the exception of a few other scholars—notably Henry A. Murray (1938)—we are little further along in our understanding of the "person-centered quartet" than was the case when James and Freud met in Worcester over a century ago.

If these fields are so central and yet have witnessed little progress, what can we expect of them in the future? I think that here we find a clue in the expansive psychologies of William James, Sigmund Freud, and Henry Murray. In one way or another, these scholars sensed an important truth: that the study of self or personality is at once *a problem of psychology and a problem of literature*. In the examples they use and in the approaches they adopt, each researcher has signaled the realization that the imaginative writer is tackling the same kinds of issues as the psychologist of personality. In James's case, of course, we have the lengthy and tortured relation with his brother Henry, as well as frequent references to other writers and to literary examples; in Freud's case, there is his reliance on the great authors of the past—Sophocles, Shakespeare, Dostoevsky—for so many of his core concepts; in Murray's case, it is his deliberate appropriation of images from literature (e.g., an American Icarus) as well as his own pioneering scholarship on Herman Melville.

But if there is a relationship between the scientific study of personality and the writer's investigation of the world of his or her characters, just what should that relationship be?

Several possibilities deserve exploration. At the very least, psychological investigators of the "person-centered quartet" ought to study works of art, including literature, with great care and test their portrayals against the claims of scientific study. Cooperative investigations among artists and psychologists could be very profitable, though the difficulty of such collaborations should not be underestimated. While the distance between psychologists and novelists might prove too great, psychologists and students of literature can each enrich one another's pursuits. Indeed, they may provide examples and "limiting cases" for one another—the psychologist's precise methods and rigor being balanced by the literary scholar's broad view and skeptical cast of mind.

The psychologist's taxonomies and frameworks need to be tested against the rich range of characters found in literature and the powerful insights about the nature of text and of reading put forth recently by literary scholars. If the schemes of psychologists prove inadequate for dealing with these more rounded examples and concepts, then they need to be reconfigured or altogether scuttled. For their part, students of literature can benefit from a study of the way in which psychologists have conceptualized the human personality, operationalized these various conceptualizations, and tested certain tantalizing hypotheses about human behavior in the experimental laboratory. Psychological and literary writers should do more than simply read one another's publications.

Here, we can take an instructive leaf from colleagues in cognitive science and neuroscience (see Essay 21). These fields have advanced in recent years in large measure because researchers reared in disparate disciplines work together shoulder-to-shoulder on problems of mutual interest. Our own modest investigations at Harvard Project Zero (see Essay 3) have for some time benefited from sustained collaborations among psychologists, artists, and experts in the systematic study of different art and literary forms. The knottiest problems in artistic analysis, such as the question of whether there might be one correct interpretation of a work of art, call for interdisciplinary investigation.

The part of psychology most likely to remain after the aforementioned cannibalizations have taken place is the study of the "person-centered quartet." Certain aspects of emotion and motivation may also elude the cognitive and neurosciences. These are topics for which psychologists may have special methods and insights, but they are equally the concern of writers and other artists, and of those who study them, like literary critics and theorists. No hard science à la physics is likely to emerge from the collaborations that I envisage. But an interesting and highly useful kind of conversation between behavioral science and the humanities is likely to occur if psychologists and individuals in the arts make common cause. This insight was not lost on our forefathers, and it has recently been reinforced in promising work by Jerome Bruner (see Essay 2), among others.

## WHITHER PSYCHOLOGISTS?

On his better days, William James was a determined optimist, but he harbored his doubts about psychology. He once declared that "there is no such thing as a science of psychology," and added that "the whole present generation (of psychologists) is predestined to become unreadable old medieval lumber, as soon as the first genuine tracks of insight are made" (Allen, 1967, p. 315). I have conveyed—indeed, confessed—my belief that, a century later, James's less optimistic vision has materialized and that it may be time to bury scientific psychology, at least as a single coherent undertaking.

Yet scientific psychologists merit praise as well.

We can rightly cherish the work of our most eminent practitioners—past and present—and the various concepts, findings, and schemes that they have developed. Whether or not psychology long endures as a self-contained field, scientists will long honor the discoveries of Donald Hebb and Karl Lashley, the concepts of identity crisis and cognitive dissonance, the laboratory procedures of psychophysics, psycholinguistics, and physiological psychology.

Even as we pay homage to our past contributors, we can participate as full members of research teams in the emerging disciplines of cognitive science, neuroscience, and, perhaps, cultural studies and developmental studies. Individuals researching in these areas will need the insights and methods of psychology—and if our colleagues do not work with us, they will only have to repeat our mistakes and reinvent our fields.

A third point: The major contribution of psychologists is to continue to tackle the most interesting problems that emerge and to follow those problems wherever they may lead. To paraphrase an old saw, "Some scientists have avoided psychology because it is too easy; but others have avoided it because it is too hard." It is in our bones—as it was in the bones of William James—to pursue the hard issues; to display an audacious curiosity about the human condition and to follow that curiosity wherever it leads.

One hundred years ago, William James's unstinting curiosity led him to physiology and thence to psychology—indeed, to founding at Harvard around 1875 the first experimental laboratory in the country and perhaps in the world. The scientists who flocked to psychology in the 20th century are as gifted a lot of scholars as any I can imagine. Perhaps today, some of those who in an earlier era would have turned to philosophy or to mental science are instead attracted to computer science, to brain science, to literature, or to literary studies. Such shifting of allegiances is understandable and appropriate. But my guess is that a healthy number of the most curious will continue to gravitate to those vexed issues that, at least in their minds, are best described as being psychological in nature.

If one of those bright students were to wander into my office in search of career advice, what would I say? I would counsel the student to look for those issues, problems, and phenomena that seem to straddle the newly emerging fields. I would have in mind those problems that partake of cognitive science and neuroscience; that lie at the boundary of the individual self and the social self; that straddle stream of consciousness as a psychological concept and stream of consciousness as a presence in literature; that raise developmental issues in a neurological context or tackle neurological issues in a developmental context; that occur at the interface of "pure cognition" and cognition as it unfolds in the school or at the working place. If psychology indeed turns out to be a field for foxes, rather than for hedgehogs—as I believe is the case—then I would try to convert psychologists into the sleekest and cleverest foxes around.

In closing, then, I find myself taking a leaf from Mark Antony. Having proposed a funeral for psychology as we know it, I have as well engaged in praise for much of what psychology has accomplished; and I have suggested that there is much productive work left for those who, for whatever reason, choose to continue to call themselves psychologists and wish to pursue the kinds of issues and questions that are traditionally considered psychological. In so doing, I believe I have been faithful to the vision of William James, a man whose intellect was far too capacious ever to be corralled into a single discipline; and who in fact thrived by alighting on a topic for a while and then moving on to another one . . . the proverbial fox, in Isaiah Berlin's figure; the Impressionist painter, in the words of G. Stanley Hall. William James's longtime colleague Théodore Flournoy put it well:

> [James's] genius is so abundant, so varied, and so little preoccupied with the appearance of contradiction that in gathering in his various utterances, one does not easily frame him into a truly harmonious whole. Indeed it is almost a question whether he himself would have been able to produce a perfectly linked and coherent system from the magnificent treasure of material which he has left us. (quoted in Allen, 1967, p. 495)

As we psychologists move to the second century of a post-Jamesian world, we could do worse than to emulate his spirit and his example.

## REFERENCES

Allen, G. W. (1967). *William James*. Viking.
Feinstein, H. (1984). *Becoming William James*. Cornell University Press.
Hughes, H. S. (1961). *Consciousness and society*. Knopf.
James, W. (1893). *Psychology*. Henry Holt.
Murray, H. A. (1938). *Explorations in personality*. Oxford University Press.

# HEIGHTS OF COGNITION: CREATIVITY

Attention to the theory of multiple intelligences drew me in other directions. Having described a number of different intelligences, and with a well-known interest in the arts, I was often asked whether there were also "multiple creativities." In truth, I did not know the answer, but I arrived at an approach that was enormously enjoyable for me. I decided to study one creative genius–presumptively—for each of the several intelligences—from Albert Einstein (logical-mathematical intelligence) to Mahatma Gandhi (interpersonal intelligence)—and to determine to what extent their creativity could be explained in terms of a single outstanding intelligence.

The answer to that question turned out to be quite complex. Virtually by definition, the poet and playwright T. S. Eliot had outstanding linguistic intelligence, while the dancer and choreographer Martha Graham stood out in terms of her bodily-kinesthetic intelligence. But my investigation revealed that each of these creative geniuses drew significantly on other intelligences; and, except for one individual in my septet (the composer Igor Stravinsky), each also exhibited clear intellectual limitations.

In this section, I also present a summary of the research reported in *Creating Minds* (Essay 23). In the quarter century since then, my colleagues and I have carried out many analyses and studies about creativity—our most recent perspective is conveyed in the review article co-authored by Emily Weinstein (Essay 24).

I have always felt that in studying a phenomenon, one should focus on unambiguous examples of that phenomenon. And so, in considering creativity, I turned naturally to individuals who beyond question would be considered highly creative. And it seemed equally natural to examine individuals who, by argument, were outstanding in at least one of the seven nominated intelligences. I report my findings in detail in my 1993 book *Creating Minds*. Herewith is a brief summary of my approach, findings, and conclusions.

# Seven Creators of the Modern Era

### THE PROBLEM

I propose that seven individuals were instrumental in developing those ideas and frameworks that we consider central to the "modern era." These individuals are the British-American poet T. S. Eliot; the physicist Albert Einstein; the painter Pablo Picasso; the composer Igor Stravinsky; the dancer and choreographer Martha Graham; the political and religious leader Mahatma Gandhi; and the psychologist Sigmund Freud.

There is nothing sacred about the contents of this list; I could very easily have added or substituted Duke Ellington, or Virginia Woolf, or V. I. Lenin, or Niels Bohr. But any account of creativity in the modern era would include at least some of these individuals; if we could explain their creativity, we would have arrived at some important understanding about the creative process.

Anyone who decides to talk about creativity has at least three points (if not strikes) against him:

1. Creativity is a huge and amorphous subject. Noam Chomsky has called it a mystery, rather than a problem, with the clear implication that scholars should not waste their time studying mysteries.
2. A good deal of the work in the psychology of creativity is worth little; the track record of researchers gives little encouragement to current workers.
3. Most individuals think of themselves as being creative, and so almost any generalization is immediately arrayed against the personal experience of the audience. There is none of the respectful distance that is routinely accorded to physicists or economists who are discussing their research.

Despite these perils, I have chosen to address this subject. My plan is to comment briefly on previous approaches to the study of creativity; to introduce a new approach, which takes off from the question "Where is creativity?"; to describe a new large-scale study that I am currently undertaking; and to report some preliminary impressions from that investigation.

Let me introduce one useful distinction: the contrast between "little c" creativity—the sort that all of us evince in our daily lives—and "big C" creativity—the kind of breakthrough that occurs only very occasionally in a society (for other Cs, see Essay 24).

Each of the individuals mentioned at the beginning of this essay can be appropriately described as having realized "big C" creativity. The decision to serve salad in the middle of a meal, rather than at the beginning or end, might be considered an instance of "little c" creativity.

Turning to previous psychological accounts of creativity, most of the work has been an effort to understand "little c" creativity. The best-known work is centered around "creativity tests," which are closely modeled after intelligence tests. On a typical creativity test, one is asked how many uses one can find for a brick or a paper clip; or how many objects one can discern in a seemingly random squiggle. In a specified amount of time, the more answers, and the more unique the answers, the more creative one is judged to be. These tests are highly reliable; if you do well on one, you will likely do well on others. But they have little if any demonstrated validity. Individuals who are creative in their work do not necessarily score better on such instruments than individuals who achieve little that is noteworthy.

Willing at least to tackle "big C" questions are the psychoanalysts. Freud proposed certain personality and motivational characteristics of individuals considered to be creative in the arts and in other domains. It may well be that creative individuals are sublimating other, more "raw" desires and that they have readier access to their unconscious. But there is no proof that the presence of a certain personality trait or motivational structure makes you creative, any more so than its absence lessens the possibility that you are creative. Furthermore, the Freudian point of view does not recognize differences across domains—characteristics of creativity are thought to apply equally across diverse subject areas.

Thanks to the cognitive revolution of recent decades, psychologists are now able to deal with ideational contents and structures as well as sheer instances of behavior. Psychologists in the cognitive tradition have devoted some attention to the processes and products of creativity. This approach focuses exclusively on what goes on within the head of the creative thinker—a reasonable tack for psychologists, but one that proves inadequate for a rounded understanding of creativity. Attempts to simulate creativity typically fall short because they presuppose the selection of a proper problem as well as a determination of the data relevant to its solution—two areas that are actually central to creative activity.

## TWO PROMISING APPROACHES

Recently, two more promising approaches have emerged. These are almost diametrically removed from each other—one highly quantitative, the other determinedly qualitative. But taken together, they represent the two most important lines of work currently being undertaken.

Dean Keith Simonton has performed "historiometric studies" (1984). In an effort to answer questions about highly creative individuals and activities, he assembles the largest database that he can and then performs statistical analyses on it. So, for instance, he can discern whether individuals who work in different domains achieve their most important work during different decades of their lives; or whether individuals who are more creative maintain a different relationship to their mentors than those who are less creative. The art in this research comes in determining which databases are relevant to answering such longstanding questions. The limitation comes in understanding the fine structure of particular creative thoughts or products.

Complementing the quantitative approach of Simonton are the efforts of the developmental psychologist Howard Gruber (1981). A student of Piaget, Gruber has been carrying out intensive case studies of highly creative individuals in the sciences, such as Charles Darwin and Piaget himself. He details the period during which such individuals arrived at their most important ideas. In the process, Gruber has discerned some characteristics that tend to permeate the lives and works of such individuals: a wide-scale "network of enterprise"; a fondness for images of wide scope; and a strong and abiding sense of purpose. The art in Gruber's work comes in attempting to capture the cognitive, affective, and personality states of the creative individual at his most critical moments. The limitation centers on the extent to which findings obtained with one individual emerge as applicable to other creative individuals as well. I believe that a happy conjunction of these two approaches would provide a tremendous boost for creativity studies.

## A PRELIMINARY DEFINITION

I begin with a definition of the individual creative person. According to this definition, which has emerged as a result of collaborative work with David Feldman (1980) and Mihaly Csikszentmihalyi (1988), a creative individual is one who regularly solves problems, fashions products, and/or poses new questions in a domain in a way that is initially considered novel but that is ultimately accepted in at least one cultural setting and/or by at least one set of experts.

This definition merits unpacking, particularly with respect to the ways in which it differs from earlier formulations. A first point has to do with *regularity*. Creative individuals are rarely if ever "flashes in the pan": they exhibit a style of living and thinking that leads to a regularity of breakthroughs, most minor, a few of epochal importance.

A second and crucial point is that creativity occurs in *domains*. That is, a creative individual is never creative across the board. Rather, creative individuals display their creativity in disciplines or crafts, usually in one domain, though occasionally, as in the case of Leonardo da Vinci, in two or even three domains. Note, however, that even Leonardo—the Renaissance genius par excellence—was hardly creative in every domain. Neither his music nor his poetry nor his relationships with other individuals are particularly noteworthy.

That creative individuals solve problems is not controversial. Indeed, most definitions restrict creativity to problem-solving, often gerrymandering behaviors in a quite extraordinary way so that they can qualify as examples of problem-solving. Quite frequently, however, creative individuals are notable less for the problems that they solve than for the *products* that they fashion—ranging from heroic symphonies to novel modes of educating young children—or for the new *questions* that they raise (note Einstein's famous query about how a light beam would appear to someone traveling at the speed of light, or John Cage's attempts to incorporate chance elements into the musical composition process).

Nearly everyone who defines creativity notes that creative behaviors are initially novel or original but that ultimately, they become accepted. If they were not initially novel, no one would consider them creative. And if they were not ultimately accepted, then they might be bizarre or anomalous but not genuinely creative.

However, my definition takes a further, more controversial turn. Per this definition, no person or work or process can be considered creative unless it is so deemed by relevant social institutions. Following Csikszentmihalyi, I term this social dimension "the field." The social institutions or fields will vary enormously from one domain to another. Thus, in the case of a domain like physics, the relevant field can be as small as a dozen knowledgeable peers; in painting, it is a mélange of critics, gallery owners, agents, and the art-loving or art-owning public; while in newspaper or magazine publishing, the field can consist of thousands or millions of purchasers. The field may operate quickly or slowly; be steadfast in its judgments or highly fickle; but in the end, the field must issue judgments about creativity, because no other set of standards can convincingly be invoked. Absent some consensus within a relevant field, the most one can say about a putatively creative individual or product is that "one can't tell."

Here, then, is a provisional definition of creativity, framed in terms of the individual. Note that this definition can easily be reworded so that it applies to a creative product (one judged as such by the field) or process (a bio logical, psychological, or sociological process that lead to a work judged as creative).

## A RESEARCH PROGRAM

Armed with this provisional definition, it is possible to lay out a program of research. As I initially formulated it some years ago, the study of creativity is inherently an interdisciplinary undertaking. Indeed, to elucidate creative persons or products, one should take into account four distinct disciplinary perspectives:

1. *The subpersonal.* Here one analyzes the creative individual in terms of neurobiological factors, particularly those of genetics, neurophysiology, and neuropsychology. So far there is little to say about creativity from the subpersonal or neurobiological perspective. But techniques already exist whereby it is possible to study the neurology of creativity *in vivo.*

2. *The personal.* Roughly speaking, this spans the psychological level (see Essays 21 and 22). Studies of creative individuals certainly should focus on cognitive aspects—the particular intellectual strengths and proclivities of the individual. Equally relevant are aspects of personality, motivation, will, and other noncognitive or less cognitive features of the human person.

3. *The impersonal.* An important part of the creative configuration is the nature of the domain in which an individual works. Adapting a term of David Feldman's, "domain" is used here as an umbrella term to cover academic disciplines as well as arts, crafts, and other pursuits that feature performances or productions of various degrees of expertise. As long as there exists a structure of knowledge and/or of performances that can be so specified that a novice can acquire expertise, it seems legitimate to talk of a domain. I use the term "impersonal" here to stress the point that at least in theory, knowledge of the domain could be presented and preserved in a nonliving entity, such as a textbook. In practice, however, most individuals come to know, and to master, a domain by virtue of an intensive apprenticeship with other individuals.

4. *The multipersonal.* As a complement to the impersonal domain, there exists the relevant field, a complex of individuals and institutions that pass judgment on the various products in a domain. Just as one must study the nature of the discipline or craft in which the creative individual works, so, too, it is important to understand the operations of the judgment-making bodies within and across domains. One might say that impersonal domain study smacks of epistemology, while the multipersonal field study relies most heavily on sociology.

## WHERE IS CREATIVITY?

I contend that the study of creativity was significantly enhanced a few years ago when Mihaly Csikszentmihalyi proposed that researchers asking, "What is creativity?" or "Who is creative?" instead tackle the question, "*Where* is creativity?" Csikszentmihalyi proposed that creativity lies not at a single locus but rather in the *dynamic interaction* among three nodes: the individual person or talent; the formal structure of knowledge in a domain; and the institutional gate-keeping mechanism, or field.

*An example*: Say that there are 1,000 budding painters at work in New York City, each with his or her peculiar strengths and styles. All of these individuals attempt some mastery of the domain of painting, as it now exists; and all of them address their work, sooner or later, to the field—the set of gallery owners, art school departments, newspaper critics, agents, and the like. Of these individuals, a few will be selected as worthy of special attention by the field; and at least today, their novelty will be a significant factor in their selection. Of these already selected individuals, at most one or two will paint in a manner that becomes so esteemed that their efforts will ultimately have some effect on the domain—on the structure of knowledge that must be mastered by the next generation of painters.

We can see here that creativity lies not in the head of the artist (or in his hand), not in the domain of practices, or in the set of judges; rather, the phenomenon of creativity can only—or at least, best—be understood by taking into account the interaction among these three nodes.

Painting, however, may seem an idiosyncratic domain. What of a contrasting domain, such as mathematics? I submit that the processes here are essentially the same. Substitute for our thousand artists an equal number of mathematicians, say topologists. Each of these students must master the domain the best they can; those who aspire to greater heights must address their proofs and their discoveries to the field—this time a set of professors, journal editors, and prize-givers. Only a few of the young topologists will stand out in terms of professorships and publications; and of these even fewer will actually affect the domain in which they work sufficiently so that the next generation of youthful topologists will encounter an altered domain.

The field is as important in mathematics as in the visual arts. It differs chiefly in that it is somewhat smaller and far more consensual in its judgments. (By an amusing coincidence, the mathematics community awards every 4 years a medal to the most gifted mathematician under the age of 40, called the Fields Medal!)

On to my study!

Between 1880 and 1930, major revolutions occurred in a number of domains, ranging from politics to science to the arts. Considered jointly, they are often termed the advent of modernism or the modern era. Reflecting the range of human talents or intelligences, as noted earlier I have studied the creative breakthroughs associated with T. S. Eliot, Albert Einstein, Pablo Picasso, Igor Stravinsky, Martha Graham, Mahatma Gandhi, and Sigmund Freud. In carrying out this study, I sought to illuminate their individual creativities, discovering those features that seem to cut across domains and persons, and those that seem restricted to particular individuals. At the same time, I hope that I may be able to say something of significance about the modern era.

Some are uncomfortable with an exercise that places a few individuals on a pedestal. Why not study people who came close but did not make it, or more ordinary individuals? Certainly my study is not an attempt to claim that one should not study the French psychologist Pierre Janet, in his day as well known as Freud; or Freud's friend Wilhelm Fliess, to whom Freud deferred during the period of his breakthrough but who turns out to have been a crackpot. Yet, again, by focusing on individuals who are considered both creative and successful, one is more likely to shed light on general issues of creative productivity.

By design, my study of each of these individuals focuses on them as persons; the domains in which they worked; and the fields which policed their domains.

## THE PERSON

Turning first to the node of the individual person, I detect two major areas of action. The first has to do with the relationships among the individual's intelligences. Creative individuals are characterized not only by outstanding intelligences (Picasso's spatial capacities, Einstein's logical-mathematical abilities), but also by *unusual configurations of intelligence* (Freud's very strong linguistic and personal understandings are unusual in a scientist; Picasso's relatively weak scholastic abilities probably reinforced his total dedication to life as a painter).

The second area of action within the person has to do with the nature of intimate relationships at the time of a creative breakthrough. Creative individuals tend to be very independent and are often quite happy to be isolates. At the time of their greatest breakthroughs, however, they characteristically need to have *an intimate to whom they can confide their thoughts.* Whether it is Freud with Wilhelm Fliess, Picasso with Georges Braque, Eliot with Ezra Pound, or Graham with Louis Horst, these vulnerable individuals gain affective as well as intellectual sustenance from an alter ego.

## THE DOMAIN

Turning next to the domain, there are again two areas of action. The first has to do with how highly structured the domain is. Academic physics and academic painting are highly structured, in the sense that there exist many levels en route to expertise and these are well delineated and agreed-upon by experts. Contemporary painting, creation of computer software or websites, and good city management are far less well-structured domains.

A second and related aspect concerns the extent to which there exists a single dominant paradigm within the domain, as opposed to two or more manifestly competing paradigms. Physics after Newton had one dominant paradigm, but in the years before Einstein, a number of less well-entrenched paradigms competed for attention. During the height of classical music, the classical sonata form became the norm; two centuries later, Stravinsky's chromatic and polytonal styles competed with Schönberg's atonal, serial style.

## THE FIELD

One facet has to do with the size of the relevant field and, in particular, the degree of hierarchy. The number of physicists at Einstein's time was relatively small, and of that number, the opinions of Hendrik Lorentz, Jules Henri Poincaré, and Max Planck were so important that they alone could virtually determine the reputation of a worker.

Another factor pertains to the degree of consensus that obtains within the domain. In certain fields at certain times the field is characterized by wide consensus, while at other times it is sharply divided. Thus, the same field that wholly embraced Stravinsky's *The Firebird* in 1910 was sharply divided about *Le Sacre du Printemps*, initially performed just three years later.

Now it might be thought that domain and field simply mirror each other, and indeed there is a strong connection between this pair of nodes. A tightly structured domain is more likely to have a hierarchically arranged field; and a domain with one dominant paradigm is more likely to achieve consensual judgments. Yet the two dimensions can and should be kept separate. At certain times a loosely structured domain (like rock and roll) may achieve a high degree of consensus (that is, regarding the merit of the Beatles); by the same token, a domain with a single dominant paradigm may exhibit little consensus, as in the case of contemporary judgments of Mozart's compositions as compared to his peers (like Antonio Salieri).

## TENSIONS ACROSS NODES

My study suggests that creativity is particularly likely to occur when a degree of tension exists between or across the three nodes. I have already suggested some of the tensions that may occur within nodes—for instance, within a set of intelligences or between two paradigms in a domain.

But perhaps more compelling are the tensions that emerge across the nodes.

Consider, for instance, the disjunction between a person and a domain. Freud's profile of intelligences was highly unusual for a scientist. Most scientists are strongest in logical-mathematical and spatial intelligences, but Freud stood out for his linguistic and personal knowledge. While he was a competent neurologist and neuroanatomist, his abilities did not mesh with his initially chosen domain of neuroscience. Once he began to study clinical psychiatry, his abilities fell into closer synchrony with the demands of the domain. Still, until he founded psychoanalysis, he was not well synchronized with the domains (and the fields) where he was working.

Picasso was an individual in tension with a field. The young and prodigious Picasso was appreciated by family and friends; and after a difficult start as a young Spaniard attempting to survive in Paris, he had begun to attract an audience and patrons as well. But Picasso deliberately turned his back on the realistic portrayals of the "blue" and "rose" periods and began to paint in a harsh and nonrepresentational way, thus inducing a strong tension between himself and the field. Ultimately, of course, his Cubist works came to be accepted—if not loved—and Picasso's highly creative output was appreciated by an ever-widening field.

There may be a tension or asynchrony between a domain and a field. In the early decades of the century, classical music veered in two directions: the harsh, rhythmic, polytonal music of Stravinsky, and the calculatedly stark and atonal 12-tone music of Schönberg. At the time it was by no means clear which of these two directions would ultimately prevail, and at various times, the field (or the fields) tilted in one direction or another; even Stravinsky occasionally composed in the 12-tone style. It now appears that 12-tone music no longer exercises a hold on the musical world; so to speak, the field has voted against serial music, while remaining more sympathetic to the option embraced by Stravinsky.

## DEVELOPMENTAL PERSPECTIVE

During the early years of development, particularly in areas where there are prodigies, one can anticipate a strong link between talents and domains; indeed, as David Feldman has demonstrated, a prodigy is defined by a strong fit. Eventually, however, a point will come when a discrepancy or asynchrony arises—between the individual and the domain, the individual and the field, or in some other internodal arena. Sometimes, as in the case of Freud, the asynchrony seems to be intrinsic to the unfamiliar line of work; in other cases, as with Picasso, the asynchrony may well be instigated by the individual himself, as a means of "raising the ante."

Looking at the broader pattern of an individual's development, the first 10 years emerge as the time when the individual attains the mastery of the domain as currently practiced. In the case of prodigies, this mastery occurs very early—by the time they were teenagers, Mozart and Picasso had already achieved the level of their masters. The crucial decade can occur much later, however: Van Gogh did not hit his stride until his thirties; Gandhi and Freud operated in domains where mastery is unlikely to come about until middle age.

In the case of the modern masters, there tends to be a strongly iconoclastic work about a decade or so after they have begun to work in earnest in their domain. It is often this "breakaway" or "breakthrough work" that first attracts attention, sometimes positively, often more controversially. Candidates for this position would be Stravinsky's The Rite of Spring, Picasso's Les Demoiselles d'Avignon, Eliot's "The Love Song of J. Alfred Prufrock," Einstein's special theory of relativity, and Freud's "Project for a Scientific Psychology."

Then, approximately a decade later, there emerges work that is more comprehensive, and more likely to exhibit integral connections to the rest of the domain. This latter work has something of the Final Statement about it. These more comprehensive works would include Stravinsky's Les Noces, Picasso's Three

*Musicians* or *Guernica,* Eliot's *The Waste Land*, Freud's *The Interpretation of Dreams*, and of course Einstein's general theory of relativity.

One can best see the asynchrony when one studies the individual at the time of his or her major breakthrough. At least in those areas where individuals are conducting *conceptual investigations,* a set of steps like the following can be discerned:

1. The individual works in an unproblematic way in his or her domain–business as usual.
2. A discrepant element emerges. The individual attempts to deal with this discrepant element by conducting some kind of local surgery. However, the discrepant element keeps reemerging and is not easily eliminated.
3. The discrepant element "spreads" and "works." Far from being an error or nuisance, the discrepant element actually provides an important and generative element. The individual seeks to understand the once-discrepant but now generative element in depth and to develop a symbol system with which to work on and integrate the element.
4. Having established that the once-discrepant element is a powerful and generative one, the individual now seeks to share these insights with a broader field. This step may involve the invention of a new symbolic form that can communicate with a wider audience; it often features miscommunication, feedback, and revision, as the individual and others attempt to test the system's limits and to promulgate the new formulation.
5. Finally, the system works sufficiently well that it is absorbed by other practitioners But sooner or later, the domain and field move in unpredictable and uncontrollable ways.

## FEATURES THAT CHARACTERIZE CREATIVE INDIVIDUALS

My studies have also suggested a number of features that can be discerned across the several creative individuals whom I have been studying.

1. *Need/inclination to campaign for oneself.* Some discoverers, like Einstein, have shown little tendency to promote themselves or their works, preferring to leave this mission to others. Others, like Freud, have been tireless promoters of their apparent achievements.
2. *Relations to rivals.* Einstein showed little interest in his rivals and was able to maintain reasonable relationships with almost all of them. Stravinsky and Picasso were far more embattled, having little good to say about anyone else.
3. *Need for close confidantes.* Nearly all of the creative individuals benefited from the reactions and support of an alter ego during the

time of their maximum isolation. But at other times, these creative individuals differed widely from one another in terms of the amount of support they needed or desired. Also, in some cases (such as Freud), the support came primarily from individuals of the same sex, whereas in other cases (such as Picasso), the support was rooted chiefly in relationships with people of the opposite sex.

4. *Role of collaboration.* In some cases—for example, poets or painters—collaboration is a luxury that need not be indulged in. Others, however, have careers in which collaboration is of the essence. Thus, composers of theatrical works, like Stravinsky, and designers of a new social movement, like Gandhi, are involved on a daily basis in complex (and often stormy) interactions with other individuals.

5. *Role of prodigiousness.* As already noted, prodigiousness is common or even expected in certain domains, such as mathematics or music. In others, however, it is rare (painting) or even unknown (clinical medicine).

6. *Ability to be creative within a given domain.* Some workers, like Picasso and Einstein, make their greatest breakthroughs in domains that have already been quite well defined. Others, however, like Gandhi and Freud, are better thought of as inventing a new domain, or perhaps radically restructuring one or more domains that had already existed.

Note some other similarities among creative individuals, not all of them expected:

1. *Combination of childlike and mature.* Creative individuals have advanced to a sophisticated and mature level in at least one domain. At the same time, however, they manage to retain a childlike quality, which enables them to pose questions in a naïve way, to see things afresh, to transgress customary boundaries. They also appear in many ways to be like an amalgam of child and adult. Possibly one reason that breakthroughs occur in early life is that it is more difficult to sustain such neoteny as the years pass.

2. *Long-term trend from the world to the self and back to the world again.* As children, these future creators are open to the experiences of the world and seem to take things in an almost effortless way. They resemble sponges, to whom virtually nothing in the world is foreign. In contrast, at the time of their major breakthrough, they often seem to be quite sealed-off, almost entirely alone. But then, following the working-through of their fundamental insight or reorientation, they once again seek to make contact with the wider world, and to locate their discovery in many contexts.

3. *The inclination to be daring and bold.* Without such audacity, and the ability to withstand criticism, it is unlikely that they could have persevered and prevailed. Interestingly, nearly all comment that they

have little sympathy for—or even little understanding of—an individual who would be content with the conventional.

4. *A certain degree of marginality.* All the creators were in some way or another marginal in their chosen domain—through place of birth, gender, religion, or some combination thereof. Perhaps more important, they felt marginal from an early age; and they continued to feel marginal even when, to others, these creators were taken to be icons of the relevant establishment.

My inquiry thus far has suggested that these asynchronies can be observed both ontogenetically, in the trajectory from apprentice to master and from the revolutionary to the synthesizer; and microgenetically, at the time when the discovery of greatest moment has been made. In the process I have also identified several features that seem to differentiate among the masters, as well as another cluster of features that so far are exemplified by all of them.

Of course, it remains a task for the future to determine whether these generalizations will hold across other creative individuals, other domains, and other times. My study does suggest one characterization of the modern era: It is a time when individuals have been able to unite some of the questions and understandings of earliest childhood—the first years of life—with the tools and understandings acquired in sophisticated mastery of a domain. Whether it is Einstein asking about traveling on a beam of light, Freud following his dreams and free associations, Gandhi achieving a goal in light of lesser physical strength, or Graham using the most elemental gestures to convey fear, hope, or triumph—all of our modern masters are at once children and sages. I doubt that one would find the same union of the toddler and the titan in other eras, but perhaps breakthroughs in other eras or epochs relate mastery to other, later stages of human childhood.

## REFERENCES

Csikszentmihalyi, M. (1988). Society, culture, and person: A systems view of creativity. In R. J. Sternberg (Ed.), *The nature of creativity* (pp. 325–339). Cambridge University Press.

Feldman, D. H. (1980). *Beyond universals to cognitive development.* Ablex.

Gardner, H. (1993). *Creating minds: On the breakthroughs that shaped our era.* Basic Books.

Gruber, H. (1981). *Darwin on man.* University of Chicago Press.

Simonton, D. K. (1984). *Genius, creativity, and leadership.* Harvard University Press.

# Creativity

## The View From Big C and the Introduction of Tiny c

*Howard Gardner and Emily Weinstein*

### BACKGROUND

As explicated in the preceding essay (23), the positing of "multiple intelligences" soon generated the question, "Are there also multiple creativities?" This led to a study of seven highly creative individuals from the 20th century and to certain conclusions, also stated there. These individuals were unambiguous examples of "big C" creativity—human creativity at the highest level—perhaps termed genius creativity.

My study in comparative biography uncovered many intriguing findings. One might equally characterize them as "hypotheses"—to tweak an old adage, the plural of "case studies" is not "data." Here are some findings that caught my fancy:

Creative individuals tend to be born in somewhat remote areas. But as soon as they can, they move to a big city, find other "young rebels," and make common cause with them.

The creators I studied came from bourgeois households. They were loved and supported by family as long as they worked hard.

Creators find or create something that is anomalous. Instead of running away from it or reverting toward "standard operating procedure," these aspiring creators become intrigued and explore further and more deeply. (See my curiosity, noted in Essay 23, about the trek from initial idea to finished work.)

The greater the anomaly, the more isolating the experience. And so creative individuals often feel that they are going insane; at such times, they especially need (and, when fortunate, receive) support from caring others.

Creative individuals gain from such support, and because they are often magnetic, charismatic personalities, other individuals like to be in their vicinity. But these supporters are often rejected when they are no longer useful—if you get too close to creative fire, you may well get singed.

Consistent with this last point, creative individuals are dedicated fully to their work—as long as they remain active in the field. To paraphrase the poet W. B. Yeats, they prioritize "perfection of the work" over "perfection of the life." As individual personalities, in their relationship to other persons, they range from remote (Einstein) to sadistic (Picasso).

Whether or not they are personally extroverted or eager for publicity, they did what was needed to bring their work to the attention of others—or were fortunate enough to have someone else who served in effect as their agent.

It takes about 10 years to develop expertise in a new area. Perhaps confirming this rule of thumb, "my" creators developed new lines of work at approximately decade intervals.

As noted, these are all hypotheses, which remain to be tested by historiometric (or, if possible, experimental) means. And when I began to speak about them, individuals often cited counterexamples.

These counterexamples have proved useful to me. As one example, the great philosopher Ludwig Wittgenstein was not born in a remote village; he grew up as a member of a Viennese family that was wealthy and well connected. A little research revealed that Wittgenstein left Vienna and went further and further afield—to a rural Austrian village; to Cambridge, England; to Ithaca, New York. Clearly, it was important to Wittgenstein to get away from his roots—and in this case, from a major metropolitan area to various comparatively remote outposts.

I am also chagrined that my list of creators was heavily skewed toward white males—Gandhi and Graham being the two exceptions. And so, in subsequent writings, I have paid attention to other creative individuals—the writer Virginia Woolf, the activist Eleanor Roosevelt—and the movement leader Martin Luther King Jr. I also studied a creative individual from an earlier era—the great composer Mozart (Gardner, 1997).

One of the rewards of the scholarly life is the opportunity to work with gifted young scholars who critique and extend one's own work. In this introduction, I introduce you to four of them.

Jin Li (1997) distinguished between *two forms of creativity*: In *vertical creativity*, the rules of the domain are quite strictly laid out and it is relatively straightforward to determine who is advancing the domain (think physics, classical ballet); in *horizontal* creativity, the rules are much looser and participants are free to move about in various directions (think postmodern literary criticism, contemporary painting).

A second student, Mia Keinanen, studied creative activity in a *horizontal domain* (modern dance). She found that rather than learning at the feet of masters (so to speak), learning was more likely to emerge as the product of regular intense interactions among peers (Keinanen & Gardner, 2004).

A third student, Katie Davis (see Essay 37) worked with me on the potentials and limitations of computer use by the "app generation" (Peppler, 2013). And the fourth, Emily Weinstein, co-authored this essay.

## THE TWENTY-FIRST CENTURY: THREE CHALLENGES

Using a terminology that I first heard from Mihaly Csikszentmihalyi, but that may have an earlier origin, I was clearly focused on "big C" creativity. As noted, the mainstream of psychological research has alighted instead on "little c" creativity. While there's no agreed upon definition, this contrast has made its way into the literature.

Someone once asked Mihaly Csikszentmihalyi whether he himself was an example of "big C" or "little c" creativity, and he quipped, "Well, I am middle C."

Whether or not that self-characterization is correct, it's clear that we need a term to describe individuals and achievements that are significant, worth noting, but that do not in themselves change a field or domain. Kaufman and Beghetto (2009) propose "pro C" to describe professional creativity—"pro C" creators are highly accomplished professional experts who have not yet (and may not ever) reach eminent status, yet who surpass the "little c" designation. In biographies, one finds the names of persons who clearly aided the "big C" creators (e.g., Louis Horst for Martha Graham, Wilhelm Fliess for Sigmund Freud, members of the "Collegial Academy" for Albert Einstein) but whom we would not consider to be highly creative.

To use a current yardstick, "big C" creators occupy a great deal of space in Wikipedia; in contrast, "middle c" creators, if mentioned at all, have little space in Wikipedia, and that space tends to focus on the person's relation to the "big C" creator.

To be sure, there are no hard-and-fast indices for the various font sizes of C. It is helpful to think of the four Cs as a continuum, with no sharp division between any two nodes. So, to use a well-known example, the composer Antonio Salieri was no Mozart, but on some criteria, he could qualify as a "big C" rather than a "middle c" creator.

Two other issues intersect in potentially intriguing ways with the size of font of the C.

My earlier work has deliberately focused on the single individual—if not the painter of lore holed up in his bohemian attic, at least the clear-cut leader of a lab or a movement. What do we know about creativity that emanates from groups—what corporations often label "Skunk Works"? Is group creativity different from individual creativity? And to take the strongly critical view, is individual creativity to some extent an illusion?

My studies all focused on individuals who did their work before the rise of computers, the Internet, the web, and ubiquitous social media. Now these forms of communication, of thinking, are ambient and perhaps determinant.

Enter another noteworthy former student (#3, for those who are counting!), Katie Davis: We asked, how does the era of the "app generation" change the context of creativity for eminent creators? And how do youths' heavy uses of social media relate to creativity? When a youth posts a Facebook entry or shares a Snapchat photo or—perhaps most revealingly—tweets a response to some event, does this count as creativity at all? (See Essay 37 as well as Essay 22 in *The Essential Howard Gardner on Education.*)

We propose to introduce a fourth "c"—tiny c—and to speculate about its relation to the larger fonts. To tackle this question, I turn to a fourth gifted former student—like the others previously mentioned, now a leading scholar— Emily Weinstein.

## EMILY

To state the obvious: Howard's academic study of creativity began well before the digital era. *Creating Minds,* published in 1993, also predates smartphones and social networking apps. In contrast, my own research focuses on the effects of digital media on users, especially young persons (Weinstein, 2014).

Today, daily digital media use is normative. Mobile phones are omnipresent, and a majority of U.S.-based adults own personal smartphones. Eighteen-to-twenty-four-year-olds check their cellphones an average of 82 times per day. Many teens and adults also actively maintain accounts on multiple social networking sites.

Social networking sites are a Web 2.0 innovation—they exemplify social, participatory platforms designed to support user-generated content. The early Internet (i.e., "Web 1.0") comprised relatively static websites created for passive consumption. With Web 2.0, networked citizens connect as content co-producers.

What are the consequences, for creativity, of a networked world? The core findings of Howard's work on creativity are almost certainly unchanged by digital technologies—that eminent creators combine two or more intelligences in distinctive ways is not a function of historical context.

However, Howard also describes collections of observations that typify the life paths of exceptional creators. Social technologies fundamentally alter the context surrounding creative individuals; one might therefore wonder whether, and to what extent, the original observations apply in a digital age. For example:

- Creative individuals tend to be born in somewhat remote areas. But as soon as they can, they move to a big city, find other "young rebels," and make common cause with them.

Enhanced connectivity is an obvious consequence of the Internet; social technologies change the nature of geographic isolation. Young creators can access the major works of their domains instantaneously regardless of where they live. The Guggenheim's online collection includes nearly 1,700 works from more than 575 artists; from the Metropolitan Museum of Art, 400,000 works are available for free, high-resolution download; and the Smithsonian's searchable database provides digital records of more than 10 million works in its collection.

Young creators also easily unite with physically distant contemporaries. Web 2.0 functionally supports affinity groups, which are often characterized by both high levels of peer support and extensive mentoring. While there is clearly less of an imperative to move to urban "creative centers," ambitious young persons do frequently move to so-called creative cities. Indeed, in the United States, and perhaps elsewhere, the difference between urban attraction and life elsewhere is greater than before. And there is no evidence that online contact substitutes adequately for in-person contacts. As has been quipped frequently, "Why do I travel long distances to conferences dedicated to the power of online learning and of virtual contacts?"

Expansive digital databases and connected learning opportunities ostensibly support "little" and "middle c" creativity. What is the consequence of new digital media (NDM) for "big C" creativity? Will future eminent creators still tend to begin life in remote areas and then, as soon as they are able, move to cities in search of affinity groups? Or do the affordances of NDM obviate the need for urban relocation?

- The greater the anomaly, the more isolating the experience. And so creative individuals often feel that they are going insane; at such times, they especially need (and, when fortunate, receive) support from caring others.

Mobile technologies and social apps expand support networks in at least two ways. First, NDM enable "tele-cocoons." Existing close ties are accessible around the clock and across the globe. Caring others are always a phone call (or text message, e-mail, video call, tweet, Snapchat, etc.) away. Mobile phones allow people to move through the world metaphorically, surrounded by tele-cocoons of personal support (Habuchi, 2005). Social apps also facilitate relational connection: When actively used for support-seeking, social networking sites activate ever-broadening networks of close and loose ties.

Second, NDM create opportunities for new forms of social support. Crisis Text Line, for example, provides "access to free, 24/7 support and information via the medium people already use and trust: text [messaging]." The text-based hotline enables immediate communication with trained volunteers. Online community groups and message boards also offer readily accessible social support. Functional barriers to connection are arguably lower than ever before.

And yet, as Sherry Turkle (2011) cautions, technology can disrupt the quality of self-reflection and social connection, distracting from and diminishing the power of deep personal relationships.

Based on my interviews with networked youth, I find that teens do routinely use social apps in the service of supportive relational connections; however, digital interactions may also exacerbate feelings of disconnection, particularly when youth already feel vulnerable or isolated. Relational exchanges on social networking sites often reflect or amplify existing social circumstances. While networked interactions can certainly comprise substantive communication, friends often demonstrate support through less personal interactions, such as "likes," "tags," and emoji-infused comments.

Howard's creators received deep, sustained support from one or two key persons. How do contemporary creators fare when they require social support to sustain themselves and their work? What role do social technologies play in creators' supportive relationships?

- Creative individuals are dedicated fully to their work—as long as they remain active in the field. To paraphrase the poet W. B. Yeats, they prioritize "perfection of the work" over "perfection of the life."

The app generation (see Essay 37) is accustomed to life on demand—immediate, efficient results across personal, relational, and professional domains. Frequent job changes among young adults arguably reflect the app mentality. A shared expectation of constant accessibility also generates pressure for rapid response, even when the quantity of digital communication becomes overwhelming. And so young adults check their cellphones on more than 80 occasions per day.

NDM open new pathways to fame (and shame), dissemination, and self-promotion. It is the age of microcelebrity—"a self-presentation technique in which people view themselves as a public persona to be consumed by others, use strategic intimacy to appeal to followers, and regard their audience as fans" (Marwick, 2015, pp. 333–334). Social technologies transfer power in the attention economy: creators or influencers can sidestep traditional gatekeepers and disseminate their own works. At the same time, creative ideas are in competition with the expansive quantities of information that "trend" in and out of view on social apps. Criticisms leveled at anything innovative, sometimes vicious, can be very discouraging—there are advantages to working in splendid obscurity, as Einstein and Freud did for many years.

In the age of self-publishing and rapid, low-cost circulation, how do different fields discern extraordinary ideas? And how do creators personally manage the opportunities for self-promotion amidst the realities of negative tweets and comments?

## TOWARD "TINY c"

Online creations pose a challenge to traditional views of creativity. Remixing—the practice of directly integrating and building on others' digital works—muddies questions about authorship and novelty. Peppler (2013) also describes the issue of defining "the field" in online spaces. Inevitably, certain ideas receive disproportionate attention and circulation among online crowds. Yet the most shared ideas are not necessarily the most creative.

Still, the potential of NDM to harness collective intelligence efficiently is potentially game-changing if one ascribes to a participatory or group view of creativity—that is, the notion that collaboration, not the individual mind, underlies creativity. Indeed, Sawyer (2007), an early investigator of group creativity, argues that all significant innovations stem from invisible collaboration webs.

Successful online collaborations exemplify what is possible with NDM, though not what is universal. For digital youth, social media creativity most often takes more ordinary forms: witty Instagram comments, playful Snapchat "masterpieces" (so-named by the platform), and clever, socially relevant tweets. The mass of these creative expressions falls below the threshold of "little c" creativity.

Over the last several years, as I conducted interviews and focus groups with dozens of networked youth, I routinely met teens who self-identify as creators. These teens leverage social-networking sites in the service of creative expression. Fifteen-year-old Lily (a pseudonym), for example, has a "meme account" that she uses to remix images with superimposed text; her goal is to produce and share original (novel) posts that will be judged by her peers as humorous, "relevant," and creative.

She shares a recent example: a picture of someone who looks dramatically exasperated and Lily's added text, "When you're almost done washing dishes and a sibling comes and puts a spoon in the sink." The post was well received by Lily's followers, as indicated by the immediate influx of likes and comments. Lily is also a painter and shares photographs of her visual artworks through a separate social media account. Throughout the day, she and her friends maintain ongoing Snapchat exchanges of digitally edited images.

Social media provides Lily with a forum to share her offline creative works, as well as a new avenue for creative expression—both of which are common experiences across her peer group. Lily's creativity on social media may certainly escalate into "little" or "middle c" creativity; however, the bulk of her social media creativity falls in the "tiny c category"—generated rapidly and forgotten as quickly. As Lily's examples reflect, "tiny c" social media creativity is also often produced in an effort for social connection, rather than as a solitary creative endeavor.

Does "tiny c" creativity go anywhere? Donning a developmental lens, Howard's studies of creators when they were young suggest that a novelist is not likely born of tweets. However, when the passion to master (and perhaps tweak) a domain is already in place, NDM provide opportunities for creativity that range from tiny to more substantial.

We began with a question about the consequence of NDM for creativity. The answer depends, in part, on one's view of the construct. Teresa Amabile (2013) describes the difference between Howard's *categorical approach* to creativity (identifying a few creative individuals) and her *continuum approach* (assuming that people are generally "capable of some degree of creativity in some domain"). If one adopts the categorical view, it makes sense to ask how creative individuals are supported and constrained by life in a digital era. To this end, we describe how NDM shift creators' surrounding contexts in potentially meaningful ways, including the provision of unprecedented access to creative works, peer groups, social support, and avenues for self-promotion.

If one adopts the continuum view, a more relevant question is whether creative expression on social apps encourages or delimits people's existing creative tendencies. We argue that everyday creative digital expressions—tweets, snaps, instas, status updates, and the like—reflect "tiny c" creativity. Whether or not "tiny c" creativity nudges the app generation toward more creative thinking overall—and we simply don't know whether it does—NDM sizably influence the context for contemporary creativity across domains.

## REFERENCES

Amabile, T. M. (2014). Big C, little c, Howard, and me. In M. Kornhaber & E. Winner (Eds.), *Mind, work, and life: A Festschrift on the occasion of Howard Gardner's 70th birthday.* Offices of Howard Gardner.

Gardner, H. (1997). *Extraordinary minds.* Basic Books.

Gardner, H., & Davis, K. (2013). *The app generation: How today's youth navigate identity, intimacy, and imagination in a digital world.* Yale University Press.

Habuchi, I. (2005). Accelerating reflexivity. In M. Ito, D. Okabe, and M. Matsuda (Eds.), *Personal, portable, pedestrian: Mobile phones in Japanese life* (pp. 165–182). MIT Press.

Kaufman, J. C., & Beghetto, R. A. (2009). Beyond big and little: The four c model of creativity. *Review of General Psychology, 13*(1), 1.

Keinanen, M., & Gardner, H. (2004). Vertical and horizontal mentoring for creativity. In R. J. Sternberg (Ed.), *Creativity: From potential to realization.* American Psychological Association.

Li, J. (1997). Creativity in horizontal and vertical domains. *Creativity Research Journal, 10*(2–3), 107–132.

Marwick, A. (2015). *You may know me from YouTube: (Micro)-celebrity in social media.* In P. D. Marshall & S. Redmond (Eds.), *A companion to celebrity* (pp. 333–350). John Wiley.

Peppler, K. (2013). Social media and creativity. In D. Lemish (Ed.), *Routledge international handbook of children, adolescents, and media* (pp. 193–200). Routledge.

Sawyer, K. (2007). *Group genius: The creative power of collaboration.* Basic Books.

Turkle, S. (2011). *Alone together: Why we expect more from technology and less from each other.* Basic Books.

Weinstein, E. (2014). The personal is political on social media: Online civic expression patterns and pathways among civically engaged youth. *International Journal of Communication, 8,* 210–233.

Weinstein, E., & James, C. (2022). *Behind their screens.* MIT Press.

Weinstein, E., & Selman, R. L. (2016). Digital stress: Adolescents' personal accounts. *New Media and Society, 18*(3), 391–409.

# LEADERSHIP

Human creativity—the focus of the previous section—has been a longtime interest of mine—dating back to graduate school, and perhaps even earlier.

In contrast, while I have also been something of a political junkie (reading newspapers each morning and magazines each evening), I had not thought nearly as much about leadership. Accordingly, when approached to write about the cognitive approach to leadership, I initially said no.

But sometimes an invitation, initially declined, circulates in one's head. In the case of leadership, I realize that—at least implicitly—I had a lot to say about what makes for a good leader. Accordingly, in the style of *Creating Minds,* I decided to conduct a set of case studies of outstanding leaders—not individuals with whose methods or goals I necessarily agreed, but ones who, in my view, had significant impact in the arenas they sought to lead.

Taking (perhaps predictably) a cognitive stance, I described leadership as an exchange between the mind of the leader, on the one hand, and the minds of his/her followers, on the other. Indeed, I see leaders as individuals who tell stories—often widely encompassing stories; and when successful, these stories can change the thinking and the behaviors of others, so-called followers.

As I've done with respect to creativity, I present here a synopsis of my findings in *Leading Minds* (1995). But even more so, my own thinking has evolved in the ensuing years. And so I also present my thoughts in 2011, 15 years after the publication of my book, and then, a decade later, a few further thoughts.

# Leadership: An Overview

Let me begin with the famous Tehran Conference of November 1943. Allied leaders Josef Stalin, Franklin Roosevelt, and Winston Churchill met for the first time and planned the fates of millions of people during the concluding phase of World War II. When most individuals think about leaders, they have in mind prototypical individuals like them.

We do not usually think of Einstein as a leader. But I want to argue that, in interesting ways, men like Einstein are leaders. What Einstein has in common with the Tehran crowd is having enormous influence on people and events. Indeed, if you think about the denouement of World War II and the 50 years afterwards, it's arguable that Einstein had as much influence as did the trio of leaders at Tehran. Not only did Einstein have the idea that led to the atomic weapons that ended World War II, but the postwar period was dominated by the nuclear threat.

It may seem even a further stretch to think of a Chinese drawing teacher in the city of Chengdu as a leader. But I am going to ask you to think about the political leaders of the world, the intellectual leaders of the world, the educational leaders of the world, and the foot soldiers of the world as being leaders in a certain sense. They are all engaged in the project of changing mental representations and transforming the way the individuals with whom they come in contact think about things and go about doing things.

## WHAT IS A LEADER?

A leader is an individual who significantly affects others—their thoughts, feelings, and behaviors. Leaders accomplish their mission in two primary ways: *They tell stories* and *they lead certain kinds of lives*.

I believe that stories are vitally important for leaders. Stories are dramatic. They feature protagonists who are trying to achieve things. There are obstacles, plenty of them.

Leaders tell stories, but it's also important to take note of the kinds of lives they lead. If leaders tell one story but lead a contradictory life, they're hypocrites. Maybe it's better to be a hypocrite than to be morally insensitive; but ultimately, if you tell one story but your own life fails to embody it, your leadership is less effective.

Seeking to be nonpartisan, I will use two quotations from political leaders. One is "I am not a crook." It's hard to say that when you are, in fact, a crook. Another is "I favor family values." It's hard to do that if, in fact, you are unfaithful, promiscuous, and/or absent. Stories don't go over in the long run if what people actually do doesn't live up to their words.

One reason why the story that Churchill told about England during World War II was so effective was because he seemed to embody the story in his own courageous, if sometimes outlandish, personal stance.

I have a cognitive view of leadership. My distinctive addition to the literature on leadership is to argue that *leadership occurs in the mind*. It occurs in the minds of leaders and followers.

*The vehicles of this mental transaction are stories*. A leader tells stories; he hopes that those stories affect what other people do. It would be easy to be a leader if everybody else's mind were a blank slate—if it were unpopulated with others' stories. Then you would just tell a story and it would be accepted or rejected on its merits.

But even by the age of 5, everybody has hundreds of stories in their mind. By the time we're adults, we have thousands of stories in our mind—some of them we're very conscious of and some we're not so conscious of.

So when a leader tries to tell a story, he or she must ask: Is anybody going to notice it? Is the story going to have any impact on anybody? If it's too remote, it's going to be ignored. We all know lecturers who are just too far from where we are, or movies that are too far from where we are, and they have no impact.

Most narratives suffer from the opposite problem. They are so familiar (like the average episode of a television series) that we have amnesia the next day; we have heard of thousands of stories like that before. It's the story that is sufficiently distant—but not so remote that we lose it—that has the potential of actually affecting us.

I see leadership as a Darwinian process where the new stories are competing with the stories already there. Often stories have rivals—what I call counter-stories. For a leader's story to have any effect, it has to slay, to nudge aside, or somehow transform or transmogrify the existing stories.

In most situations, identity stories are at a premium. People are very interested in who they are individually and corporately. One of the principal things that leaders tell us is who we are and what we're aspiring to. We pay particular attention to those *identity stories* because they convey issues that are important to us.

British Prime Minister Margaret Thatcher was clearly a leader. She was one of the most effective leaders of the second half of the 20th century. But she provoked ambivalent feelings in many of us. She's a very good example of somebody who was an excellent storyteller and who embodies her story. When she came into power in 1979, her story was this:

> England had lost its way. Socialism was a bad thing. England was once a great
> power—a powerful, imperial, fair nation of shopkeepers doing their own stuff,

working hard, getting ahead. Instead, the fat Socialists, with the tacit tolerance of my own Tory party, had formed a post-World War II consensus. England was becoming a minor power, not working very hard; unions were too powerful and nationalizing all of the industries.

Thatcher was going to set England on her right course.

Thatcher not only had a story that England was ready to hear; she also appeared to embody that story. She was a self-made person. Her father was not wealthy. She lived above the grocery store that he ran. She went to school and got two advanced degrees on her own. She was the first woman "shadow leader" of her party; of course, the first woman prime minister in British history; and the longest-running prime minister of the century. She was seen as courageous in the Falkland Islands War and after the bombing in Brighton, where she almost lost her life.

So, in sum, she looked like the kind of person she was talking about. She wasn't simply a good actor or pretender; she seemed to really *walk the talk*. I think one of the reasons why General and Secretary of State Colin Powell appealed to the American people was that he seemed to live what he talked about. He wasn't just somebody who had a good speechwriter or spin controller. That's what's distinctive about a cognitive view of leadership.

As a leader, you really have to engage the counterstories, show where they are inadequate, what's wrong, say, with the Labor position in postwar England. You have to make people feel that the position you are taking in your story—and in the life that you are ostensibly leading—can serve as a model—that it is really a much more effective way of conducting business and achieving laudable goals.

The leaders whom many of us admire—whether it's Martin Luther King Jr. or Nelson Mandela—are people who really showed us how our versions had inadequacies. They showed us not just by being persuasive but by living a certain kind of life. Gandhi led a life that made it very clear to his followers what kind of human beings they ought to strive to become.

## A STUDY OF LEADERS

With Emma Laskin, I carried out a study that was reported in *Leading Minds*. We undertook biographical studies of eleven 20th-century leaders, along with shorter examinations of leaders during World War II. They didn't just tell familiar stories (I would call such a person a "manager" of an organization). We were interested in people who changed the way other people are, not people who simply reinforced current habits or ideas.

The leaders in the study were volitional in the sense that they were persuasive—they did not compel agreement through force. My sample was also skewed toward individuals who tell *inclusionary* stories—stories that involve many people in their "we." (Actually, it turns out that being inclusionary is not necessarily advantageous for a leader; sometimes it is more effective to be exclusionary, to characterize individuals as being outside of your chosen circle.)

Let me mention the individuals that were studied. Two people led scholarly domains: Margaret Mead, the anthropologist, and Robert Oppenheimer, the physicist who actually put Einstein's ideas into practice by leading the Manhattan Project. I call these people *indirect leaders* who sought to become *direct leaders*. These people led by the force of their ideas.

Then I studied people who led circumscribed institutions, among them Robert Maynard Hutchins, a very effective leader of the University of Chicago in the 1930s and 1940s. One of the ways that we know that he was effective is that 50 years— now, over 70 years!—after Hutchins stopped being president, the University of Chicago still bears the shadow of his conception and his contributions.

I studied people who led what I call the classical estates. Alfred Sloan was the extremely successful head of General Motors. When he led an organization, he made it a point to travel around the country and confer with the people who ran the different franchises. George Marshall led an even bigger estate called the U.S. Army. There were 8.5 million soldiers under Marshall in World War II. Pope John XXIII led a group of about one billion people called the Catholic Church.

Note that these people inherited organizations or institutions that already existed.

Leadership is entirely different when you don't have a group and you have to create one. The dynamics between people who already have expectations, who already know a certain story, differ from the dynamics that characterize people who need to create new stories. Creating a new story is a much, much more difficult problem, and it's harder to sustain your enterprise once you are no longer on the scene.

I examined two people who sought to lead beyond national boundaries. Mahatma Gandhi, the architect of Indian independence, had enormous impact in the United States, South Africa, China, Russia, and so on. And I studied a person who has had enormous influence on the 20th century and whose influence has continued to grow, Jean Monnet. Monnet was the French economist who had the idea of the European Union, the Common Market; he worked for 60 years to try to get it started, and as of this writing he appears to be crowned with success.

As sort of a "control group" for my more intensive studies, I also examined 10 leaders from World War II, including Hitler, Mussolini, Roosevelt, Stalin, and Chiang Kai-shek.

## THE INTELLIGENCES OF A LEADER

Let me relay just a few findings from this book-length study.

A first finding concerns the intelligences of the leader. Given what I've said, there are three intelligences that are important for a leader. One is *linguistic;*

leaders have to tell stories. Occasionally they can do it in another symbol system, but by and large they do it in words, orally, although sometimes in writing.

They need *interpersonal intelligence*—an understanding of other people—because leaders try to get people to change the way they are. You can't do that if you don't have considerable understanding of other people.

I think leaders also traffic in *existential intelligence* because they help us understand who we—their potential followers—are. They answer these existential questions of what are our goals, what are the obstacles, why are we here, why do we live, why do we die?

One interesting thing is that leaders don't particularly require *logical-mathematical intelligence*. In fact, there's a study that shows a negative correlation between understanding of economics and being a good leader. Interestingly, Ronald Reagan was consistently underestimated throughout his life because he wasn't a particularly logical character. He was, in fact, a brilliant storyteller. In Washington most people are lawyers and they are logical; many of them thought that Reagan was a fool. But Reagan wasn't trafficking in that particular commodity. It wasn't important for his effectiveness as a leader.

## LEADERS TAKE RISKS, DEFY AUTHORITY

An early marker of leadership is the capacity to challenge: When they're young, leaders take risks and defy authority. They are willing to confront people in authority, not necessarily in an in-your-face way, but more to say, "I've analyzed the situation and I see it somewhat differently, and this is how I see it." If they were too in-your-face, they would get killed or marginalized. The trick is to do it in a way where you're taken seriously.

A textbook example of that is General George Marshall, who was not a particularly aggressive character. In fact, as military people go, he was rather taciturn. The first time he met General John Pershing during World War I and the first time he met Franklin Roosevelt before World War II, he told them off in public—an incredible thing to do. In fact, Secretary of the Treasury Henry Morgenthau said to Marshall, "Nice knowing you, George," because you didn't tell President Roosevelt off in public the first time you met him and still survive.

But, in fact, in both cases, Marshall was hired immediately afterward to be chief of staff. So there is something about *the way* that these people can take risks and confront authority that marks a future leader.

Now, to anticipate my story a bit, I said to myself, "Well, people like Einstein or Oppenheimer or Margaret Mead didn't confront authority, and so maybe they're not leaders." But, in fact, I didn't take my own analysis seriously enough.

So-called *indirect leaders* tell their stories through the work they do. What you'll find with creative people who are *indirect leaders* is that they are telling off the people in authority, but they do it through their work. Einstein didn't have to read the riot act to his professors. He simply rendered them irrelevant with his own analysis of time and space. As did Picasso with his paintings, and Schönberg and Stravinsky and the Beatles with their music. So there is a defying of authority and risk-taking among indirect leaders, but it doesn't occur face-to-face; it occurs through the actual iconoclastic work that they do.

Leadership is very different if you inherit an organization with stories and norms than if you're creating a movement. If you don't have an organization, a lot of your energy has to be devoted toward creating one with some stability. Many people think that the Internet will be the locus of new power in organizations. I'm actually skeptical because, to put it bluntly, it's too easy to drop out of the Internet. People would have to establish norms where you couldn't just drop in and drop out, and, of course, that goes against the spirit of the Internet.

Leaders need *opportunities for reflection*. The interesting thing is that indirect leaders or creators—the Einstein types—really spend about 90% of their time reflecting and about 10% of their time interacting with the rest of the world and making sure that what they're thinking isn't crazy.

In contrast, consider people who are direct leaders, like school superintendents or principals. Ninety percent of their time is spent with other people, and that's probably necessary, but that last 10% of the time is really crucial. If you don't have time to reflect and see the big picture, you are very vulnerable to costly mistakes. I worry about people who spend all their nights going out, who don't have vacations, including time for thinking while on holiday.

## LEADERS TRY AND FAIL . . . AND THEN TRY AGAIN, QUITE POSSIBLY WITH A DIFFERENT TACK

Finally, leaders are not people who succeed all the time. In fact, most leaders fail more than the rest of us because they are very ambitious and far-reaching and because they're trying to change things. Their lives are punctuated with failures. Where they differ, at least from me if not from you, is in the way they *frame* the things that don't go well.

If you're dealing with a group of experts, a homogeneous group that knows a lot about your topic, then you can convey a fairly sophisticated story. When economists or the Bar Association meets, the stories that are conveyed are pretty complicated.

However, most leaders are facing heterogeneous groups, collections of people who, even though they may include experts, encompass individuals who are certainly not experts in the same thing. With heterogeneous groups, stories need to be very simple because they are basically addressed to the unschooled mind.

I got interested in leadership because Gandhi was so different from the other creative people whom I'd studied; he was dealing principally with an "unschooled mind" (see Essay 16 in *The Essential Howard Gardner on Education*) and a heterogeneous group.

I had thought that it was always good to have the big "we" and to have a very inclusionary story. But there's a risk, in fact. The broader the "we," the more that people who are your core constituents don't feel special anymore. They may feel alienated. They may feel they've been neglected. Margaret Thatcher, in fact, got very far by being exclusionary. Her favorite question was, "Is he or she one of us?," with the obvious implication: if not, "off with his or her head." She was very oppositional and found that to be a very productive stance.

It's no accident that leaders like Gandhi and Yitzhak Rabin were assassinated not by people from outside their circle but by people from inside their core groups, individuals who felt that the designated leader was not being a good enough Hindu or a good enough Jew.

Similarly, Martin Luther King Jr.'s loss of influence occurred when he stopped having African Americans from the South as his core constituency, moved toward poor people in general, moved his operations to the North, and shifted to political issues like the Vietnam War. Then people who could speak more directly to his earlier constituencies—Stokely Carmichael, H. Rap Brown, and Malcolm X—became the more effective exclusionary kinds of leaders.

## A LEADER AND AUTHENTICITY

Finally, the most effective stories grow out of the life of the leader. They have an authenticity because the narrative represents experiences that the leader has gone through and has grappled with. The opposite side of the coin occurs when you try to tell a story that somebody else has written for you. In the long run, it won't be convincing because your own life won't embody that story. In fact, I think the best stories grow out of lifelong embodiments. This doesn't mean that you have to be like everybody else. But it does mean that not only does your story have to be authentic, it also has to touch on certain aspects of your constituency even if, in superficial aspects, those constituents are very different from you and from one another.

Years ago, I was in a meeting where many of the leaders of the world were in attendance, many influential people like Bill Gates, co-founder and original CEO of Microsoft. Everybody wanted to see Bill Gates because he's admired so widely nowadays. But there was one leader who, when he walked into the meeting, everybody stood up for automatically. This was Nelson Mandela.

Most people in that room had little in common with Nelson Mandela except that he's a human being, and they believed they could connect to him on the human level. When Nelson Mandela got sworn in as president of South Africa, he had his jailer sitting in the first row. That's such a Gandhian touch.

Going back to the Bible and probably to prehistory, the leaders conveyed stories. Incidentally, because Moses wasn't able to speak very well, he had to contract with a storyteller—his brother Aaron—but most of us don't have that luxury. Charles Darwin did. Darwin had a colleague named Thomas Huxley who could go on the hustings, and that turned out to be very important.

Do leaders embody their story? Christ is one seamless embodiment. Do they have an audience? And how do they create and maintain one, particularly if their story is jarring? What is their organizational and institutional base? How do they maintain it and keep it from regressing into being a mere organization rather than a place with productive tension that is at least in part a movement?

Leaders have to tell stories that are simple because the groups are heterogeneous. Members won't understand a complicated story.

The opposite situation: We may understand the technical aspects very well and, like most academics, tell stories that are too complicated. All leaders need to wrestle with these forces, which I think are more pressing now than they would have been when I studied these issues at the end of the 20th century.

## INDIRECT LEADERS: CREATIVE MINDS

Let me return to Einstein, Mead, Stravinsky, Freud, and other people whom we would call *creators* (see Essay 23). I call them *indirect leaders*. Both indirect and direct leaders are trying to influence people. But indirect leaders do it through works that they create, what I would call *symbolic products*: pieces of paper with squiggles on them, whether they're drawings or scores of music or poetry or scientific theories or big fat books about intelligence. That's the means whereby the influence is wrought or not wrought. Don't think that indirect leaders are less powerful than direct leaders.

The economist John Maynard Keynes had a shrewd, indeed profound insight about the power of indirect leaders. As he quipped, "Practical men who believe themselves to be quite exempt from any intellectual influence, are usually the slaves of some defunct economist" (Keynes, 1936).

Indirect leaders are very powerful, but you may not know them. It doesn't matter to science what Einstein looked like, what he talked about to his friends, or even, to be frank, whether he loved his wife. What matters is the quality of the symbols that he created and whether they influenced other physicists. Indirect leaders deal with other experts in their domain, other physicists, other composers. They lead by creating products that affect the work of other people in their domains, which in turn change the domains: Their embodiment is how they go about doing their work. People learn about how Einstein thought, and that's very important for people who want to continue to be physicists.

And the same thing happens to anybody who is an indirect kind of leader. Because you're dealing with experts, the stories or the products that they deal with can be quite sophisticated. Therefore, creators contrast with direct leaders, those who work with groups that are heterogeneous, that need to be spoken to directly. Because direct leaders are dealing with unschooled minds, their stories have to be simple. The simple stories drive out the complex stories. In a struggle, the simple stories always win initially.

A word about the sector of education: Educators are leaders, and leaders are educators. There are two primary reasons for this statement. One is more descriptive, one's more normative.

Educators are in the business of transforming the mental representations of their students. Otherwise, education is a waste of time. Educators shouldn't be occupying space unless their goal is to change mental representations, and they can do that in many ways, including telling stories and, more important, by the kind of lives they lead. My one-sentence recipe for desirable character education: Hang around parents and teachers who have good character.

The second and more normative connection: Both educators and leaders face and confront the unschooled mind: leaders because they're dealing with heterogeneous groups, educators because the youngsters they are confronted with are unschooled, or there wouldn't be any need to educate them. The job of educators is to school unschooled minds, to transform the mental representations. It's very hard to change those mental representations. It's very hard to school the unschooled mind. So those are nontrivial connections between educators and leaders. (See Essays 16, 25, and 26 in *The Essential Howard Gardner on Education*.)

## THREE LESSONS

What can we learn from extraordinary people—creators, direct leaders, and indirect leaders?

There are three lessons: reflecting, leveraging, and framing.

### Reflecting

Reflecting means spending a lot of time thinking about what you're doing and why and building in time for such reflection. These people all do a lot of reflecting. Even people like Churchill built time in their lives for reflecting. In fact, Churchill used to take baths all morning, and he wasn't just sleeping, he was thinking about things.

Not reflecting is an abiding evil, or at least vice, of our time. Reflection is not a habit we develop sufficiently in kids. I think our 5-year-olds reflect as much as our 16-year-olds, and that is a troublesome state of affairs.

### Leveraging

Leveraging is figuring out what you're good at, pushing it, and not worrying too much about what you're not good at. If you want to achieve something notable, figuring out your competitive advantage, what the niche is that you're better able to fill than other people, is an adaptive thing to do. In fact, the word *adaptive* is probably appropriate because it's very Darwinian.

Crying about spilled milk, things that you're not that good at, is not a very good use of time. You can try to get better at things you're not good at for fun and profit, but it's not really where your contributions are going to inhere. Extraordinary people spend a lot of time leveraging.

### Framing

The third concept, framing, grows out of the fact that people who are extraordinary have lots of bad things happen to them; they experience lots of failures. It comes with the territory. Many of us, including myself, when failing are prone either to give up or to ignore it. Both of these reactions are bad ideas.

Framing means that when something hasn't gone well, recognize that it didn't go well but try to learn from it. Try to treat it as a positive experience, even as a moral lesson. Analyze it, dissect it, figure out what you might do differently.

## DEFEAT IS AN OPPORTUNITY

The most inspiring phrase I came across in my whole study was from Jean Monnet, the founder of the European Union. He said, "I regard every defeat as an opportunity." I've been literally stunned by how frequently that line appears in the study that I've done of extraordinary people.

This turns out to be a real marker of extraordinary people. They convert apparent failures and defeats into learning opportunities. I also think it's a terribly important thing for parents, teachers, superintendents, and principals to do. When things don't go well, they need to recognize that in their own lives (and then in the lives of the community for which they're responsible), they need to try to treat these setbacks as a learning experience. I love Esther Dyson's phrase, "Make new mistakes." It's terrible to make no mistakes. It's good to make new mistakes and to learn from them. Jean Monnet and Mahatma Gandhi—these are my heroes.

## REFERENCE

Keynes, J. M. (1936). *The general theory of employment, interest and money*. Palgrave Macmillan.

# On Good Leadership

## Reflections on *Leading Minds* After Three Decades

Today (writing in 2024), I still believe this account in general. But events of the past years have given me considerable pause.

By 2010, some wrinkles or challenges to my account were already becoming clear. In a new edition of *Leading Minds* published the following year, I reflected on "Leadership in the Era of Truthiness, Twaddle, and Twitter":

> *No leader today can afford to ignore this powerful trio: The ease of promulgating false statements; the detritus that permeates the blogosphere; and the prominence of the ad line and the gag line. Indeed the challenge to the leader is to counter these forces when they are inimical to his or her goals and to put forth a powerful counter-story—one that highlights truth against truthiness, clarity against twaddle, and a developed and substantiated story as opposed to a twitter-length teaser. As I write these lines, US president Barack Obama clearly understands these challenges; but it is uncertain whether he—or, indeed any thoughtful leader capable of complex thought—can be heard and understood above the din. (Gardner, 2011, p. xii)*

Of course, the threats to authentic stories, compellingly told and actually lived, have been exemplified by the persona and behavior of President Donald Trump. But I don't want to focus unduly on Trump because we hear similar contrived stories and encounter analogous faux embodiments around the world. Consider the words and actions of contemporary leaders—Bolsonaro (Brazil), Duterte (Philippines), Orban (Hungary), Erdogan (Turkey), Xi (China), Putin (Russia) . . . and the list could be easily extended.

Nor are these threats limited to the early 21st century. My cautionary words of 2010–2011 could (and perhaps should) have been applicable in the 1930s—in the years leading up to World War II—and no doubt in earlier eras as well. It has long been tempting for leaders to create powerful myths—posing as heroic loners, arrayed against the forces of evil—and to persuade an impressively sizable cohort of followers that what they say is true and that, as a consequence, their edicts should be followed. (Niccolò Machiavelli would not have been surprised.) And as I pointed out in *Leading Minds*, a simple or even simplistic story all too often prevails over one that may be more accurate and more appropriate and more truthful, but also more complex.

## AN APPROACH TO GOOD LEADERSHIP

In recent years, as part of what we call The Good Project (thegoodproject.org), my colleagues and I have shifted our focus from what makes for an *effective* leader to what makes for a *good* leader. And in this line of research, we have identified the three key features of a good leader:

> *Excellence:* The good leader knows the field in which he occupies an influential role, keeps up with developments, and draws on his knowledge appropriately.

> *Engagement:* The good leader cares about her work, finds it meaningful, and looks forward to carrying it out effectively—even at times when conditions are not favorable.

> *Ethics:* The good leader ponders the ethical implications of contemplated words and actions, strives to do the right thing, reflects on unexpected consequences, and seeks to do better the next time.

It's not always easy to determine whether someone is a good leader. With respect to excellence, many leaders rely on previous knowledge and/or do not know how to proceed when conditions change significantly. With respect to engagement, one may well be deeply engaged in carrying out work that is compromised or even malevolent (see Shakespeare's histories and tragedies).

My colleagues and I have been particularly concerned with the ethical dimension. In almost any position of leadership, there are certain well-established norms and rules that can and should be followed. In following such norms, one does not need to exercise one's ethical muscles.

The ethical test occurs when challenges arise for which the standard procedures are not adequate or appropriate, and when the leader recognizes this conundrum. I've termed this recognition the ethical "aha." Indeed, if you don't recognize and then attempt to deal with the new situation, there is no possibility of an "aha" nor, accordingly, for pursuing a better course of action.

In such challenging situations, leaders need to reflect on intentions or motives, the means at their disposal, and the necessity of dealing with the consequences of the actions that they undertake or choose deliberately *not* to undertake. Accordingly, we judge the "goodness" of leaders in terms of recognition, action, consequences, and lessons learned. And, of course, the cycle continues throughout the tenure of the leader.

Here's a rough metric that one can apply in an evaluation of whether leaders qualify as good leaders:

1. They seek to determine the truth and tell the truth; and when they have made errors, they admit it and try to make amends.

2. They recognize the existing norms and abide by them, or are willing to challenge them openly and bear the consequences (which might entail civil disobedience).
3. When the norms are not adequate, or new issues arise, they publicly acknowledge this situation—I call this awareness the ethical "aha."
4. They articulate and ponder the dilemma—they don't claim to have all the answers.
5. They search for the best input—expert and political—including advice from a "team of rivals."
6. They make a decision openly, anticipate the consequences, and are poised to change course as necessary and to revisit the consequences of actions taken or not taken.
7. They indicate their willingness to repeat this cycle and, ultimately, help to bring about a new or revised norm of ethical awareness and reflectiveness with respect to the conditions with which they have been dealing.

In revisiting my initial conception of leadership, I have emphasized the need to take into account a fast-changing landscape, and the pressures to master the most popular media of communication. Also, I no longer take for granted a democratic society with clear standards of right and wrong and with a faith in the importance of the truth. Rather, I have focused on what it means to be a *good* leader and on the properties and processes that a good leader needs when faced with challenging dilemmas. In a phrase, we don't need more leaders—we need *better* ones; and we need to help those with leadership potential to deploy their gifts in pro-social ways.

### REFERENCE

Gardner, H. (2011). *Leading minds*. Basic Books.

# Changing Minds
## 80/20 and Five Rs

If leaders change the minds of others, an obvious question arises: In what ways, and with which tools, do leaders typically work when they are trying to change the minds of others? Carrying out case studies, and further reflection on this issue, led me to the following proposition: There are *seven principal ways in which leaders (and the rest of us as well) change minds*. I still believe that these seven "levers" are the means at the disposal of aspiring leaders.

But looking back on this study, it is also to some extent a period piece, because at the time—around 2000—I had little awareness of the emerging power of digital media—social media in particular—and the enormous impact that they exert on social discourse and behavior. These media parcellate society such that individuals naturally gravitate to persons and sites with which they are already in sympathy. This fact makes it far more difficult to change minds—and also, as is now widely appreciated, make almost all of us far less willing to engage in discourse with those with whom we believe (rightly or wrongly) we have little in common.

Also, as a scholar, living primarily in a scholarly community, I had far more faith in the role of the first two levers—reason and research—in bringing about mind change. At the time of this writing, I believe that emotional factors—what I call resonance—are far more powerful, and I would emphasize even more the importance of understanding the reasons that individuals resist change.

Still, we cannot, we should not, give up efforts at rational discourse and at changing minds through reason and research. To do so, we need to retain what I call a "disinterested stance"—an effort to figure out what is really going on in a situation, and to be willing to give up even our long and deeply held convictions, if they cannot be supported. I discuss my views of a disinterested stance in the professions and how it is severely challenged in the digital era.

### MENTAL REPRESENTATIONS: THE 80/20 PRINCIPLE

Consider a change of mind that many individuals have experienced over the years. From early childhood, most of us have operated under the following assumption: When confronted with a task, we should work as hard as we can and devote approximately equal time to each part of that task. According to this 50/50 principle, if we have to learn a piece of music, or master a new game, or fill out some role at home or at work, we should spread our effort equally across the various components.

Now consider another perspective on this issue. Early in the last century, the Italian economist and sociologist Vilfredo Pareto proposed what has come to be known as the "80/20 principle" or rule. As explained by Richard Koch in a charming book, *The 80/20 Principle* (1998), one can in general accomplish most of what one wants—perhaps up to 80% of the target—with only a relatively modest amount of effort, perhaps only 20% of expected effort (see Figure 27.1). It is important to be judicious about where one places one's efforts, and to be alert to "tipping points" that abruptly bring a goal within (or beyond) reach. Conversely, one should avoid the natural temptation to inject equal amounts of energy into every part of a task, problem, project, or hobby; or to lavish equal amounts of attention on every employee, every friend, or every worry.

Why should anyone change his or her mind from operating under the 50/50 principle to believing Pareto's apparently counterintuitive proposition? Let's consider some concrete instances.

Studies show that in most businesses, about 80% of the profits come from 20% of the products. Clearly it makes sense to devote attention and retain the profitable products while dropping the losers. In most businesses, the top workers produce far more than their share of profits; thus one should reward the top workers while trying to ease out the unproductive (and unprofitable) ones. Complementing this notion (and with a nod to pessimists), 80% of the trouble in a workforce characteristically comes from a small number of troublemakers— who, unless they are relatives of the boss, should promptly be excised from the company.

The same ratio applies to customers: The best customers or clients account for most of our successes, while the vast majority of clientele contribute little to our bottom line. With respect to almost any product or project, one can accomplish the basic specification and goals with only about one-fifth of the customary effort; nearly all remaining efforts are expended simply to reach personal satisfaction or to satisfy our own obsessive streak. In each case, one must ask: Do we truly desire such perfection? What are the opportunity costs of devoting significant energy to just one of a raft of possible endeavors?

The 80/20 principle even crops up in current events. According to *The New York Times,* 20% of baggage screeners at airports account for 80% of the mistakes. Responding to this need, an aviation expert named Michael Cantor designed a simple perceptual task that "screens out" the least able screeners.

By now, even if you have never heard of this principle, you have probably gotten the gist of it (maybe even 80% thereof!). You'll have a sense of whether this is familiar territory for you ("Pareto was just talking about cutting our losses"), or whether it represents a genuinely new way of thinking about things ("I'm going right down to the director of human resources and see how we can get rid of the most moribund 20% of our team"). You'll probably have some questions—for example, is it always 80/20? How do you know which 20% to focus on? Do we really want our pilots, our surgeons, our scientists, or our artists to practice 80/20 triage? And if you are a bit irreverent, you may ask: "How could someone named Koch write a 300-page book on the 80/20 principle?" Quick answer: It's a good read.

The 80/20 principle is perhaps best described as a concept. Human beings think in concepts, and our minds are stocked with concepts of all sorts some tangible (the concept of furniture, the concept of a meal), others far more abstract (the concept of democracy, or gravity, or the gross national product). As concepts become more familiar, they often seem more concrete, and one becomes able to think of them in almost the same way that one thinks of something one can touch or taste. On a first encounter, the 80/20 principle may seem abstract and elusive, but after you have used it for a while, and played with it in various contexts, this principle can become as familiar and cuddly as an old stuffed teddy bear.

Moreover, the more familiar a concept, the easier it is to think of in various ways. Which brings me to an important point: Presenting multiple versions of the same concept can be an extremely powerful way to change someone's mind. So far, we have described the 80/20 principle in words and numbers—two common external marks (readily perceptible symbols that stand for concepts.) But the principle need not be confined to linguistic or numerical symbolization—and it is the possibility of expression in a variety of symbolic forms that often facilitates mind-changing. See Figure 27.1, one graphic depiction of the principle.

These various ways of thinking about Pareto's principle brings us to an important point about mental representations: They have both a content and a form, or format. The content is the basic idea that is contained in the

**Figure 27.1. The 80/20 Principle**

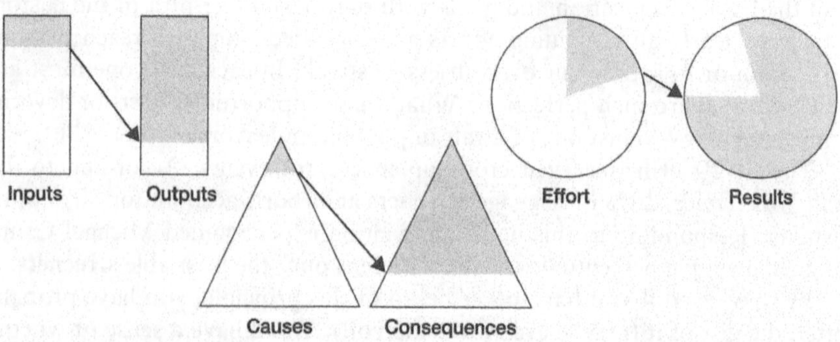

From Richard Koch, *The 80/20 Principle* (New York: Currency/Doubleday, 1988). Reprinted by permission of Random House.

representation—what linguists would call the semantics of the message. The form or format is the particular language or system of symbols or notation in which the content is presented.

I have identified seven factors—sometimes I'll call them levers—that could be at work in these and all cases of a change of mind. As a running example, I draw on an individual—author Nicholson Baker—who is trying to decide how to furnish his apartment (Baker, 1982). As it happens, each factor conveniently begins with the letters "re."

## 1. REASON

Especially among those who deem themselves to be educated, the use of reason figures heavily in matters of belief. A rational approach involves identifying relevant factors, weighing each in turn, and making an overall assessment. Reason can involve sheer logic, the use of analogies, or the creation of taxonomies. Encountering the 80/20 principle for the first time, an individual guided by rationality would attempt to identify all of the relevant considerations and weigh them proportionately; such a procedure would help him to determine whether to subscribe to the 80/20 principle in general, and whether to apply it in a particular instance. Faced with a decision about how to furnish his apartment, Baker might come up with a list of pros and cons before reaching a final judgment.

## 2. RESEARCH

Complementing the use of argument is the collection of relevant data. Those with scientific training can proceed in a systematic manner, perhaps even using statistical tests to verify—or cast doubt on—promising trends. But research need not be formal; it need only entail the identification of relevant cases and a considered judgment about whether they warrant a change of mind. A manager who has been exposed to the 80/20 principle might study whether its claims—for example, those about sales figures or employee difficulty—are borne out on her watch. Naturally, to the extent that the research confirms the 80/20 principle, it is more likely to guide behavior and thought. Author Baker might conduct formal or informal research on the costs of various materials and on the opinions of those who would be likely to visit his newly furnished apartment.

## 3. RESONANCE

Reason and research appeal to the cognitive aspects of the human mind; resonance denotes the affective component. A view, idea, or perspective resonates to the extent that it feels right to an individual, seems to fit the current situation, and convinces the person that further considerations are superfluous. It is possible, of course, that resonance follows on the use of reason and/or research; but it is equally possible that the fit occurs at an unconscious level, and that the resonant intuition is in conflict with the more sober considerations of Rational Man or Woman.

Resonance often comes about because one feels a "relation" to a mind-changer, finds that person "reliable," or "respects" that person—three additional "re" terms. To the extent that the move to forklifts and backhoes resonates for him. Mr. Baker may proceed with the redecoration. To the extent that 80/20 comes to feel like a better approach than 60/40 or 50/50, it is likely to be adopted by a decision-maker in an organization.

## BY THE WAY: RHETORIC

I note that *rhetoric* is a principal vehicle for changing minds. Rhetoric may rely on many components: In most cases, rhetoric works best when it encompasses tight logic, draws on relevant research, and resonates with an audience (perhaps in light of some of the other "re" factors just mentioned). Too bad rhetoric has that "h" as a second letter!

## 4. REPRESENTATIONAL REDESCRIPTIONS (REDESCRIPTIONS FOR SHORT)

The next factor sounds technical, but the point is simple enough. A change of mind becomes convincing to the extent that it lends itself to presentation in a number of different forms, with these forms reinforcing one another. I noted previously that it is possible to present the 80/20 principle in a number of different linguistic, numerical, and graphic ways; by the same token, as I've shown, a group of individuals can readily come up with different mental versions of Mr. Baker's proposed furnishings. Particularly when it comes to matters of instruction—be it in an elementary school classroom or a managerial workshop—the potential for expressing the desired lesson in many compatible formats is crucial.

## 5. RESOURCES AND REWARDS

In the cases discussed so far, the possibilities for mind-changing lie within the reach of any individual whose mind is open. Sometimes, however, mind-changing is more likely to occur when considerable resources can be drawn on. Suppose that a philanthropist decides to bankroll a nonprofit agency that is willing to adopt the 80/20 principle in all of its activities. The balance might tip. Or suppose that an enterprising interior decorator decides to give Mr. Baker all of the materials that he needs at cost, or even for free. Again, the opportunity to redecorate at little cost may tip the balance.

Looked at from the psychological perspective, the provision of resources is an instance of positive reinforcement—another "re" term. Individuals are being rewarded for one course of behavior and thought rather than the other. Ultimately, however, unless the new course of thought is concordant with other criteria—reason, resonance, and research, for example—it is unlikely to last beyond the provision of resources.

Two other factors also influence mind-changing, but in ways somewhat different from the five outlined so far.

## 6. REAL-WORLD EVENTS

Sometimes an event occurs in the broader society that affects many individuals, not just those who are contemplating a mind change. Examples are wars, hurricanes, terrorist attacks, economic depressions—or, on a more positive side, eras of peace and prosperity, the availability of medical treatments that prevent illness or lengthen life, or the ascendancy of a benign leader or group or political party. Legislation could implement policies like the 80/20 rule. It is conceivable that a law could be passed (say, in Singapore) that would permit or mandate special bonuses for workers who are unusually productive, while deducting wages from those who are unproductive. Such legislation would push businesses toward adopting an 80/20 principle, even in areas where they had been following a more conventional 50/50 course. Turning to our running example, an economic depression could nullify Mr. Baker's plans for refurnishing his apartment, whereas a long era of prosperity could make it easier. (He could even purchase a second "experimental" apartment!)

## 7. RESISTANCES

The six factors identified so far can all aid in an effort to change minds. However, the existence of only facilitating factors is unrealistic.

Consider the major paradox of mind-changing. While it is easy and natural to change one's mind during the first years of life, it becomes difficult to alter one's mind as the years pass. The reason, in brief, is that we develop strong views and perspectives that are increasingly resistant to change. Any effort to understand the changing of minds must take into account the power of various resistances. Such resistances make it easy, second nature, for most of us to revert to the 50/50 principle, even after the advantages of the 80/20 principle have been set forth convincingly.

Mr. Baker, for example, might elect to retain his current pattern of apartment furnishing, even when reason, resonances, rewards, and the like issue their

Circean song. The hassle of moving and the possibility that he or others might become disenchanted with the extra backhoes and forklifts might overpower several pushes toward the new furnishings.

\*      \*      \*

I've now introduced the seven factors that play crucial roles in mind-changing. As we look at individual cases of successful or unsuccessful changes of mind, we can see these various factors at work in distinctive ways. For the moment, I will only say that a mind change is most likely to come about when the first six factors operate in consort *and* the resistances are relatively weak. Conversely, mind-changing is unlikely to come about when the resistances are strong, and the other factors do not point strongly in one direction.

## REFERENCES

Baker, N. 1982. Changes of mind. In *The sizes of thought: Essays and other lumber.* Random House.

Koch, R. (1998). *The 80/20 principle: The secret of achieving more with less.* Currency.

# POSITIVE USES OF MIND: INTRODUCTION TO GOOD WORK

I have mentioned some of the theoretical and empirical offspring of MI theory. But perhaps the biggest influence on me came unexpectedly from a personal experience

In 1993, I learned that the education department of one of the Australian states was drawing on MI theory. So far, so good—I had even worked on a television program in Australia, *Lift Off*, that was inspired by MI theory.

But when I learned about the specifics of the educational program, I was dismayed. As part of the program—which may have been well intentioned—the various racial and ethnic groups in Australia were described, along with the intelligences that they allegedly had in abundance, and the ones that they apparently lacked!

I was crestfallen. Until that point, I had been pleased by the interest shown by educators in my work. I had thought that my obligation was merely to announce and explicate my findings and then move on. I had *not* anticipated to what extent these ideas could be *misused*—indeed, used in a way that I found harmful and destructive.

It so happened that two of my close colleagues in psychology—Mihaly (Mike) Csikszentmihalyi (an expert on motivation, most associated with the idea of "flow") and William (Bill) Damon (an expert in moral development, most associated with the idea of "a sense of purpose")—had also seen some of their psychological concepts misused.

The three of us had the valuable opportunity to spend a year together at the Center for Advanced Study in the Behavioral Sciences, located on the campus of Stanford University. As a result of our consultation and collaboration there, we arrived at what would ultimately be a 10-year project, soon dubbed The Good Work Project.

209

We used this invaluable opportunity to ponder a big question: Can scholars have some influence on how their ideas are used, and, if so, how can they best navigate this usage in a positive way? Out of this year's reflection emerged a project to which each of us devoted much of the ensuing decades. And our initial point of departure was on the work of *professionals*—individuals who have chosen a career in which it is possible to have some control over how one acts and how one deals with the consequences.

While the project involved many researchers on several campuses, the question that we were investigating can be succinctly stated: "How do individuals who want to do good work succeed or fail at a time when things are changing very quickly; technology is transformative and often disruptive; and market forces are powerful and not easily countered?"

In an attempt to answer this question, we spoke to over 1,200 Americans across nine different professional fields. And we identified the three essential components of good work—the so-called "three Es": *excellence* (work needs to be carried out according to the highest standards of the profession); *engaging* (work needs to be personally meaningful to the worker); and *ethical* (work needs to be carried out a way that is not self- serving, that follows the desirable norms of the calling, and that addresses the wants and needs of clients and the wider community).

Now under the succinct title "The Good Project," the enterprise has continued until the third decade of the 21st century. It has yielded many insights and much writing—indeed, 10 books and a few hundred articles, research reports, and blogs. (Not that quantity necessarily ensures quality!) Some of the work is clearly educational, and it is presented in the companion volume. Parts of the work focus on citizenship, which can also be described in terms of the three Es.

My own interest has been directed toward what it means to carry out good work in the professions. I consider the established professions—medicine, law, teaching, auditing—as well as the aspiring professions—journalism, computing—to be among the most important and precious of human inventions. It is hard to imagine contemporary society existing without competent doctors, nurses, auditors, and teachers. And yet, for internal reasons (sheer malpractice on the part of some practitioners) as well as external pressures (increasingly powerful AI programs), the professions are under sharp critique attack in our times: it is by no means clear *which* callings will survive and *under what terms*. There is also the question of the emergence of new professions, particularly in the digital realm, and how they will be realized, recognized, practiced, and passed on. Accordingly, my own writing focuses on this facet of good work.

In the following set of articles, I lay out various dimensions of what we now call The Good Project.

In two Tanner Lectures (Essays 28 and 29), I described the origins of the project; key elements of the framework that emerged; and the first lines of work that we initiated.

# Tanner Lecture #1

## What Is Good Work?

Most scholars like to keep the divisions between fields as clear as possible. Philosophers are charged with defining a terrain, making key distinctions, poring over ambiguous cases, and placing items in the optimal, most air-tight arrangement. For his skills along these lines, Aristotle was long honored as The Philosopher.

Psychologists test hypotheses about human and animal behavior and thought. These scholars strive to develop clean, unambiguous tasks: When administered properly, these tasks should allow comparisons between treatment and control groups and, ultimately, identification of causal mechanisms and chains.

Policymakers may draw on several fields. Their task is to define a problem within society, review the available data, commission new data if possible, then recommend a course of action. Policymakers may be quite promiscuous in the lines that they cross. Like scholars, policymakers are expected to keep their personal predilections in check. But this requirement of strict disinterestedness proves difficult for most of us to honor, and scholars are no exception.

In this and the succeeding essay, I will be speaking about the Good Work Project, an academic endeavor that challenges the strictures to which I've referred. (Note, 2024: The project has been renamed "The Good Project.") A long-term collaboration with psychologists Mihaly Csikszentmihalyi and William Damon, the Good Work Project started as qualitative social science—a set of interviews of prominent leaders about the work that they do. Most of the analyses we carried out were qualitative, but as the subject population grew, it became possible to pursue quantitative analyses and even to test some hypotheses. The findings themselves proved intriguing, and, indeed, we continue to mine the data for answers to questions about work.

Somewhere along the way, the tenor of the project changed. Part of the change reflected broader trends in the culture of our time; part came about because of the often unsettling nature of the findings themselves. And so, what had begun as a typically academic research project gradually transmogrified into an examination of current policies and practices in the United States and perhaps elsewhere. And before much longer, my colleagues and I were no longer just addressing policy issues; we had actually become actors—to be sure, in a modest way—seeking to bring about changes that we believed were desirable. Our interventions comprise the topic of the second part of these lectures.

All scholarly work must begin somewhere, and ours began in psychology, particularly in the subfield called developmental psychology. Developmental psychology is often thought to be the study of children, but it is more precisely described as an examination of how human cognitive and behavioral structures become increasingly complex, more differentiated, and, ultimately, better integrated.

The prototypical (and, in my view, the greatest) developmental psychologist was Jean Piaget (1896–1980; see Essays 1, 5, and 12).

As befits any great scholarly pioneer, Piaget first dominated the field of study, then was subjected to searching (as well as inappropriate) criticisms, and now is recognized for his pathbreaking but imperfect work. Most of Piaget's research focused on cognitive development—he laid out the developmental sequences outlining children's increasing mastery of senses of time, space, number, and physical objects.

While Piaget was most interested in the fundamentals of reason, especially critical and logical thought, he also explored the moral sense. In a classic monograph, *The Moral Judgment of the Child* (1932/1965), Piaget pointed out that young children focus on the *consequences* of an action—if you break five teacups rather than a single one, you are more culpable, independent of the reasons for the breakage.

Toward middle childhood, children come to recognize a crucial distinction. One is culpable to the extent that an action is intentional, and far less so if an action is innocent, accidental, or well intentioned. Thus a 10-year-old recognizes that the child who is trying to help his mother prepare the table and breaks five teacups is less culpable, less worthy of punishment, than the child who, in the course of stealing cookies, accidentally breaks a single teacup.

An American psychologist, Lawrence Kohlberg (1982), extended Piaget's work significantly. In the 1950s and 1960s, Kohlberg posed a series of dilemmas to young persons, all of whom happened to be boys. Like Piaget, Kohlberg was less interested in the moral judgments per se (right/wrong, good/bad) than in the *reasoning* that underlay and gave rise to judgments.

Ultimately, in tried-and-true developmental tradition, Kohlberg laid out a series of six stages of moral judgment through which, by hypothesis, all human beings pass. Young children are absolutists—A is right, B is wrong, might is right, the weak shall inherit nothing. Children in later childhood are rule followers—they want to do *exactly* what the society stipulates in the way it stipulates. Adolescents and adults at higher moral stages do not consider rules as sacrosanct. They are prepared to disobey them if the rules seem unjust, geared up to make and justify their own decisions, and willing to accept the consequences of civil disobedience. We see the end state of moral development in such paragons as Martin Luther King Jr., Nelson Mandela, and Mahatma Gandhi.

In a post-Piaget, post-Kohlberg era, we affirm that moral judgments are important, but they do not always predict moral behavior. Some sophisticated judges behave immorally, while individuals at so-called lower stages can and do exhibit moral courage. We also recognize a wider variety of spheres: social conventions (drive on the right side of the road) are not the same as moral strictures (drive so as to protect the lives of passengers); individuals may behave differently in their personal lives than in their professional lives; one can injure individuals directly or carry out behaviors that violate ethical codes.

In 1994–1995, Mihaly Csikszentmihalyi, William Damon, and I spent a year on sabbatical at the Center for Advanced Study in the Behavioral Sciences at Stanford University. This was a time when markets were highly valued in the country—we might say it was a quintessential neoliberal era.

The three researchers had nothing against markets per se—indeed, we were—and remained—beneficiaries of the market. Yet we were far less confident than were many others about the "genius of the market" or about its inherent "righting mechanisms," particularly when it came to taking care of the less able or the less fortunate. We were not at all sure that societies could function properly if medical care, legal protection, education, and accounting were all left strictly to market forces. At the same time, as students and aficionados of creativity, we did not want to impose constraints on individuals or society unless they were exceedingly well advised.

From discussions along these lines, we began to formulate a project that we tentatively christened the "humane creativity" project. The question was this: Is it possible to have individuals, institutions, and societies that are at once *creative* and *innovative*, and yet at the same time are also *humane*, providing for those who cannot fend for themselves? In a sense that I did not realize at the time, we were probing Adam Smith's conviction that markets could be consonant with moral sentiments—though we thought that this consonance was something to be achieved, rather than a fundamental operating assumption of a civilized society.

As empirical social scientists, we immediately pondered how to collect data relevant to our guiding question. Rather than thinking through the issues, in the manner of an armchair philosopher, or moving directly to regulations or laws, as a policymaker might do, we instead elected to *interview leaders in different sectors* to see whether, and if so, how, it might prove possible to combine creativity with humaneness.

Happily, the William and Flora Hewlett Foundation gave us some funds to launch the project. In due course, we received support from several other sources and carried out a far more ambitious project than we could have envisioned during our California year.

Nonscholars might have difficulty believing that it took us *nearly 5 years* to come up with a good name for the project and to formulate the precise question that we were tackling. We moved away from the study of creativity per se, and toward an examination of major professions in America. We did so because it is easier to study humane conduct when there are clear guidelines for ethical and nonethical behavior than when ethics are left largely or even totally up to the individual practitioner. Put concretely, doctors and lawyers are enjoined to proceed according to specific guidelines and can sacrifice their reputations and lose their licenses if they behave in an unethical manner. In contrast, workers in business and the arts essentially have no comparable restrictions. They are essentially free to do what they want (including behaving ethically or unethically) as long as they do not run afoul of the law

As for the name, we decided to call ourselves the Good Work® Project. (As you can see, we even registered the trademark, something I have never done before or since.) Initially, we defined Good Work as work that is *excellent* in technical quality and carried out in an *ethical* manner. Subsequently, we added a third criterion and an alliterative third E—that the work be *engaging* and personally meaningful. The word "good" thus draws on three separate connotations: good quality, feels good, and embodies the good pole of morality. We represent this conception via the accompanying image (Figure 28.1).

For whatever reason, people prefer the term "good work" to the term "humane creativity," though many—perhaps even most—individuals refer to our endeavor incorrectly as the Good Works project.

On to the question that we posed as researchers. We asked: "How do individuals who want to do good work—work that is excellent in quality, personally engaging and meaningful, and carried out ethically—succeed or fail at times when things are changing very quickly; our senses of time and space are being altered by technology (such as the Internet and the World Wide Web); market forces are very powerful; and there are not comparable forces—religious, ideological, communitarian—that can counter or moderate the market forces?"

Note that the question does not critique markets per se. But the question— and the questioners—are forces that can temper the Darwinian, dog-eat-dog quality that all too often characterizes a completely laissez-faire system.

**Figure 28.1. The Triple Helix of Good Work: Excellence, Ethics, and Engagement**

Excellence Ethics Engagement

So, armed with a name and a question, we created a comprehensive questionnaire that contained 60 questions with follow-ups, and sampled eight different topics: goals and purposes; beliefs and values; the work process; positive and negative pressures; formative influences; training of the next generation; community and family; and ethical standards.

Importantly, we sought nominations of individuals in various domains who were thought to be good workers. The nominators included both experts in the field and individuals who themselves had been nominated—the so-called "snowball" method of amassing subjects. We had no independent proof that each of our subjects was actually a good worker; and indeed, in a few cases, a subject was subsequently revealed to have behaved in an unethical or even illegal way. Nonetheless, in the aggregate, we were confident that our subjects were reliable informants about good work, whether or not each exemplified good work in every particular.

Every research subject was administered a version of our research protocol. Interviews averaged one and a half hours, with some of them running substantially longer. On the average, a transcription of an interview ran to 30 single-spaced pages, and often 40 or even 50 pages. The interviews are best described as *semi-structured*. That is, we carried them out in a conversational way, letting the subject direct us comfortably from one sphere to another, and yet made sure that at least the eight major topical areas were covered in each case.

In addition, in most domains, we administered more quantitative measures. As examples, we posted ethical dilemmas to subjects and noted how they resolved these conundrums—for example, should a journalist insist on interviewing a bereaved person when the family asks for privacy; should a government-supported researcher secure a patent on a discovery in her own name? In nearly all cases, we also administered a so-called Q sort, in which subjects were asked to indicate which personal values (e.g., integrity, fame, responsibility, financial security) were most important and which were least important to them.

Initially, because of our own interests and the availability of funding, we focused on two so-called *domains*—genetics and journalism. The coupling proved fortunate for both conceptual and empirical reasons. Conceptually, geneticists have the most to say about our bodies, while journalists affect what is in our minds. As we quipped, we were studying the guardians of our "genes" and our "memes." Empirically, the coupling was fortunate because it turned out that the two domains differ dramatically from each other in terms of the experiences and reflections of the practitioners.

In the end, during the period 1996–2004, we interviewed over 1,200 subjects drawn from nine different professional domains: genetics, journalism, theater, precollegiate education, higher education, law, medicine, business, and philanthropy. Ironically, we were unable to raise money to study law or medicine, the two most prominent professions (eventually we did so, though on a shoestring). We had to be creative, as well as ethical, if we were to survey these two domains.

In sharp contrast, we were awarded more money to study philanthropy than any other domain. Perhaps there is a moral (or even an immoral!) to that story. (See Essay 31 on "The Lonely Profession.")

We surveyed quite widely: promising young individuals just entering these domains, individuals in midlife, veterans, and individuals we nicknamed "trustees"—persons no longer as active as they had once been but with an impressive concern about the health of the domain. The demography of the subjects paralleled that of the domain; there were many more female teachers than female geneticists. When funding permitted, we secured a truly national survey, but in most cases, we focused on the coasts, where our respective universities were located. All subjects lived in the United States, though quite a few had been born abroad. These points need to be kept in mind when one considers the generalizability of the findings.

We collected massive amounts of information, much of which can properly be called data. Some of these data are unashamedly quantitative—how many subjects were male or African American, mentioned God explicitly, considered switching occupations, and so forth.

But most of the data are better described as qualitative. That is, we picked a topic of interest—for instance, whether and, if so, in which way a subject exhibited a sense of responsibility. In such a case, we reviewed the transcripts for discussions of responsibility, created a coding system, and then categorized the responses of subjects in terms of their differing senses and loci of responsibility.

It would take a lifetime to analyze the data exhaustively. Indeed, our book *Responsibility at Work* (Gardner, 2007) focused on just one of our several dozen questions: "To whom or what do you feel responsible?"

Before moving to the theoretical framework that we ultimately arrived at, and a sampling of our results, a few comments about the "feel" of this kind of research—the sort of information that rarely gets into the academic journals.

First of all, this was a large-scale collaboration. In addition to the three principal investigators (PIs), at least 50 other researchers participated in the research during the decade-long project. Two of the PIs moved to new universities during the period, thus necessitating the setting-up of new research offices and the hiring and training of new staff.

Funding for the research was never completely in hand: Many researchers did not know the source of their salaries from one year to the next, and in a few instances, researchers had to be let go because of lack of funds. The cycles of grant renewals do not necessarily coincide with availability of research subjects or payroll demands. Considerable resourcefulness was required to keep the project afloat. Ultimately, we had two dozen funders.

That said, the three PIs and the rest of the team worked together well, and with remarkably few jolts and hitches. At least part of this smooth operation was due to the longstanding personal connections among the PIs and their respect for and trust of one another. The advent of computer file-sharing and excellent

Internet connections made possible collaboration that would have been far more difficult even a decade before.

I would add, however, that the researchers who joined the team also believed in the mission of the project. This conviction resulted both in a "can-do" attitude and in the capacity to solve problems rather than harp on them or sweep them under the rug. I'd go so far as to label them "good workers."

Recruitment of subjects was not always easy. We were requesting from subjects a considerable amount of time and a considerable degree of candor, with essentially no compensation in return (subjects were each offered a book authored by one of the PIs). In general, we had a high success rate in recruiting subjects, and once recruited, nearly all subjects cooperated throughout the study. In cases where we felt that a particular subject was crucial to the study but also elusive, we made use of personal contacts and were usually successful in carrying out the interview. (One very busy, very famous subject gave me only a half-hour but managed to go through the entire protocol in that abbreviated period!) We respected the subjects' privacy and assured them that any and all remarks that they wished to be off the record would be so treated.

It is worth nothing that subjects often commented on how useful the interviews were—the sessions had given them a rare and valued chance *to reflect on their lives*. And subjects' memories of the conditions of study could also shift over time. More than one subject requested anonymity during the session, only to ask us, at a later time, "When will the book about my work be published?!"

Over the course of the project, a useful theoretical framework emerged. Inspired by an analytic framework that had been developed by Csikszentmihalyi (1988) in his earlier studies of creativity, we posited the interactions in work of four crucial factors (Figure 28.2):

A. The domain, discipline, or profession under study. Domains are cultural inventions: sets of beliefs, practices, and values—often encoded in a symbol system—that are developed over time and passed on from one generation of workers to the next. As conveyed by the ancient but still relevant Hippocratic oath, medicine constitutes a powerful domain, one that in the West has served as a model for other domains and professions.

B. The field or social ambit within which the domain is situated. Surrounding any profession is an ensemble of training institutions, gatekeepers, prizes, sorting mechanisms, and punitive agencies—all devised by those having an interest in the domain to regulate its operation. Within medicine, the field consists of such assorted entities as medical texts, medical school admission departments, internships and residencies, and the bodies that issue and/or withdraw licenses. Whereas the sum of domains constitutes the culture, the sum of fields constitutes the society.

C. The individual is the locus of all work. Whatever the rewards and strictures of the ambient society, in the end it is the individual—with her beliefs,

**Figure 28.2. A Graphic Rendition of the Principal Elements of Good Work**

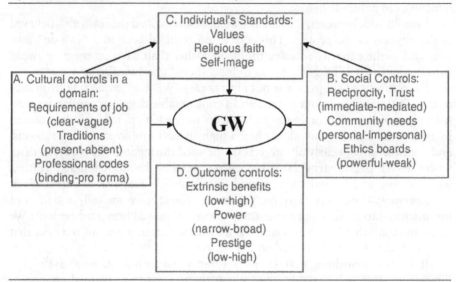

values, goals, motivations, personality, temperament, fears, ambivalences, and so on—that takes a stance vis-à-vis the precepts of the domain and is ultimately placed, honored, or chastised by the field. With respect to our example, the domain of medicine in the United States consists (at the time this essay was written) of the approximately 850,000 individuals who are empowered to practice medicine.

D. The control mechanisms of the society. Individuals from their chosen domains located in societal fields are inevitably subjected to broader forces within the society: the opportunities available, the prevalent rewarding and punishing mechanisms, the messages prevalent on the street and in the media. In the United States of the early 21st century, powerful messages about the hegemony of the market and the need to succeed financially exert potent pressures on the practice and the practitioners of medicine. Such forces can give rise to physicians who participate in large HMOs or who risk hanging up a single shingle; who work in consort with drug companies or who travel to a developing country to help control an outbreak of a disease; who practice concierge medicine or campaign for single-payer health plans.

All four of these forces are always present. The ways in which they operate and interact determine the likelihood of good work.

Let me turn to some findings.

A. *Alignment.* Our first and in many ways fundamental finding grows directly out of the application of our framework. In our search for individuals who exemplify good work and institutions that are hospitable to good work, we identified a crucial factor—that of *alignment* as opposed to *misalignment* or *nonalignment* of crucial factors. Briefly put, good work is easier to carry out in instances when all of the major stakeholders involved with the domain want the same thing. Conversely, good work proves elusive in cases where the various stakeholders espouse different and often conflicting goals.

As it happens, the first two domains studied—genetics and journalism—provided starkly instructive contrasting cases. In the United States at the end of the 20th century, genetics turned out to be a well-aligned domain. The principal stakeholders—the scientists, their funders, the core values of science, the current social institutions—all wanted the same thing: basic and applied discoveries that would lead to better health and longer lives. Geneticists were seen as key to this goal, and so all roads were wide open to their dedicated pursuit of science. None of the geneticists we studied actively considered leaving the domain—most were eager to get up in the morning, go to work, and discover new and valued knowledge, techniques, or products.

Journalism presented a dramatically contrasting case. At the start of the new millennium, American journalism turned out to be a massively misaligned profession. (This statement remains true a quarter of a century later.) Many reporters had joined the profession for idealistic reasons—they wanted to investigate the most important stories, report them fairly, bring truth to power, and do so in a way that was respectable rather than sleazy.

But most journalists felt stymied at nearly every point. The tastes of the public called for sensation, not substance. News outlets belonged not to individual families (as they had earlier in the century) but to large multinational corporations that were fixated on ever-greater profits—and, it appeared, *only* in profits. For them, selling newspapers or TV news programs was no different than selling chairs, corsets, or cigarettes. Investigative reporting—an indispensable accoutrement of a truly democratic society—was frowned upon: it was expensive, politically controversial, and—worst of all—might embarrass the advertisers or the publishers.

No wonder that fully a third of our subjects considered leaving journalism, and a majority felt that the domain was moving in the wrong direction. Alas, subsequent trends have only confirmed the continuing misalignment of journalism, with concomitant damage to its principal values, and little sense of what may be ahead. (Note, 2024: And all this was before the advent of powerful social media and sites that purport to present the news—and that make the practice of the profession of journalism even more challenging.)

To be sure, alignment does not guarantee good work, nor does misalignment preclude it. Even in the most aligned of fields, some sinners will be inexpert workers, or feel alienated, or cut corners. And even in the least aligned fields, some saints will pursue work that is excellent, engaging, and ethical. Indeed, some good workers seem *highly motivated by misalignment*—scholar Noam Chomsky and lawyer Ralph Nader come to mind. Yet, all things considered, it is far easier on practitioners bent on carrying out good work if they happen to have chosen professions that are well aligned in their time.

B. *Differences Across Professions.* The finding about alignment cuts across professions. But each profession also has its own particular topography, a fact that counsels caution about generalizations across the workplace. In the case of education, and particularly higher education, many decisions occur at the institutional level, rather than at the individual level. And so, in our study, we elected to begin by identifying institutions—rather than individuals—that themselves seemed to embody good work. Our measure of good work in higher education began with an examination of the extent to which priorities expressed by faculty, students, and the institutions' mission statements were aligned (Note, 2024: The large national study of higher education undertaken subsequently by Wendy Fischman and me [2022] draws directly on this approach.)

Another example: In the case of law, we regularly encountered an institution that was rarely mentioned in any other domain—that of the partnership. And so we ended up studying what makes for a robust partnership and which forces can weaken it or even trigger its collapse.

The economic basis of a profession might seem an important factor in the quality of its work. And, indeed, we found that precollegiate teachers' struggles to make ends meet threatened the quality and/or the longevity of their work. Yet economic security does not guarantee work quality or satisfaction. We happened to study theater and philanthropy at about the same time. We had every expectation that theater would be a domain under stress, because of the difficulty of finding well-paying (or even paying!) work; whereas in philanthropy, with no need to raise funds and with relatively high salaries, we expected the work to be unproblematic.

And yet we discovered almost the opposite picture. A large number of grant-givers reported feelings of uncertainty and anxiety, because there was no reliable way of determining the quality of their work (see Essay 31). In contrast, despite economic uncertainties, individuals in the theater reported remarkable engagement with and passion for their work. Perhaps only someone who really loved the art form would endure such an uncertain vocation. (Note, 2024: This was before the COVID-19 pandemic, which posed particular challenges for performing artists.)

Finally, the quality of work is at risk when individuals enter a field in search of great fortunes, rather than because they cherish the work per se. Those who entered medicine in the hope of becoming wealthy were much less likely to embrace their work than those who valued the treatment of patients or the discovery of new knowledge.

C. *Individual and Group Differences.* Somewhat to our surprise, group differences were not particularly salient. On most measures, respondents within a profession aggregated with their peers. Put differently, the differences we encountered were as likely, or even more likely, to reflect the choice of profession (and the current atmosphere within that profession) rather than one's membership within that profession.

That said, some individual or group differences emerged. On the average, women and minorities reported less of a sense of "agency" or "control of work conditions" than did men and members of majority groups. Other than that, gender differences were not salient. Asked to whom or what they felt responsible, a large number of African Americans cited their ancestors who had sacrificed so much. To a lesser extent, this sense of fealty to one's forerunners was found among Jewish and Asian subjects, while, intriguingly, it was virtually never mentioned by white Anglo-Saxon subjects.

By far the most striking individual differences were found between older and younger subjects. Older subjects generally endorsed the values of their profession, honored those who adhered to them, spoke appreciatively of their mentors, and reported their own efforts to behave in an ethical manner.

In contrast, our younger subjects reported a much more mixed picture. They less often cited mentors and more frequently lamented the absence of mentors or other heroic figures. Often they focused instead on a *tormentor* or *anti-mentor*. And when pressed for role models, they were more likely to combine features of different persons—a practice that we've dubbed *frag-mentoring*.

Like their senior counterparts, the young workers could identify the qualities of good work, and they expressed their admiration for good workers. But a large minority of these subjects stated that they could not or would not at this time carry out work that is fully ethical. Their stated reasons: They were ambitious and wanted to succeed; they knew or suspected that their peers were cutting corners or executing compromised work; and they were simply unprepared to sacrifice their own chances for success by behaving in a way that was more admirable than that exhibited by the peers with whom they were competing. Some stated this position reluctantly or apologetically, while others were quite matter-of-fact or even defiant—"that's just the way the world is." This finding deeply disturbed us and prompted efforts at intervention (see Fischman et al., 2004).

D. *Responsibility.* When we arrayed our subjects by age, we also discovered an ever widening circle of responsibility. The youngest subjects are most likely to express responsibility to those immediately around them: *family, friends, their own personal agenda*. As they become more firmly established, they enlarge the circle of responsibility; it now comes to include the *institution* for which they work and, at least sometimes, the particular *domain* in which they are working. Scientists talk about the importance of adhering to the codes of ethical science, while lawyers speak about their responsibility to the courts and to the pursuit of justice. Among our oldest veterans, we occasionally encounter individuals who see themselves as responsible for the continuing health of the domain. As noted, we've dubbed these wise persons the *trustees* of the domain.

A useful division in the conceptualization of responsibility emerged across domains. Individuals in professions like medicine or teaching speak primarily of their responsibility to the *individuals* whom they serve on a daily basis—patients and students, respectively. In contrast, those in other professions express a responsibility to a more abstract conceptualization. Thus, journalists speak of their responsibility to a broader public, to democracy, to the ideal of a free press; grant-makers cite a sphere—justice, the environment, the eradication of poverty or disease.

We certainly would not say that one form of responsibility is more valuable or more to be cherished than its complement. Both are needed. What is not clear is whether individuals with different senses of responsibility (to person vs. to career path) are drawn to different professions, or whether the formative processes and ambient atmosphere of the profession themselves inculcate different priorities of responsibility.

One surprising finding was the lack of invocation of responsibility to God or to a higher power. In one of the most religious countries on the planet, only a handful of subjects explicitly mentioned God. About one-seventh of the subjects cited a spiritual element in their work; this could include the feeling that they were "called" to their work or a sense of mystery or oneness with the universe while they were in the throes of work. Quite a few subjects mentioned the importance of values acquired during early childhood and they often credited religious training, whether or not they considered themselves religious at the present time.

The relatively low religious response rate in our study could have been due to the preponderance of subjects from coastal regions; to a reluctance on the part of professionals to speak explicitly of their religious commitments to a social science researcher; or, as I suspect, to a widespread bifurcation between work life and one's personal religious persona. Even in the United States, professionals may render onto Caesar those things that belong to his realm.

Clearly, we have not solved the issue of what determines good or compromised work in our society at the present time. But I hope at least to have given some hints of the likely factors, a way of thinking about the findings, and a few frameworks for considering your own work and the work of those whom you know and cherish.

In the accompanying essay (Essay 29), I turn to issues of policy and values as I address a pressing question: How can we cultivate good work in the young?

## REFERENCES

Csikszentmihalyi, M. (1988). Society, culture and person: A systems view of creativity. In R. Sternberg (ed.), *The nature of creativity: Contemporary psychological perspectives*. Cambridge University Press.
Fischman, W., & Gardner, H. (2022). *The real world of college*. MIT Press.
Fischman, W., Solomon, B., Greenspan, D., & Gardner, H. (2004). *Making good: How young people cope with moral dilemmas at work*. Harvard University Press.
Gardner, H. (Ed.). (2007). *Responsibility at work: How leading professionals act (or don't act) responsibly*. Jossey-Bass.
Kohlberg, L. (1982). *The psychology of moral development* (Vol. 2). Harper & Row.
Piaget, J. (1932). *The moral judgment of the child*. Free Press.

# Tanner Lecture #2

## Achieving Good Work in Turbulent Times

Perhaps the most striking and the most disturbing result to emerge from our study concerns the testimony provided by young workers—individuals ranging in age from 15 to 30 or thereabouts. These young subjects could readily distinguish good work from compromised work; many admired good workers; and some—in a most impressive manner—strove to carry out good work themselves.

But a significant minority of young workers rejected good work as an immediate goal. As they put it, sometimes regretfully, sometimes defiantly, they wanted to succeed themselves as they perceive it; many of their competitors were cutting corners; and so they found it necessary to suspend their ethical sensibility, at least until such time as they themselves had gained the desiderata of fame, power, wealth—the sweet smells of success.

One can assume that these young people are articulating their genuine beliefs. There would be little reason for these young persons—the proverbial "best and brightest"—to portray themselves deliberately in a less favorable light. Moreover, many other pieces of data—some from our own subsequent studies and interventions, some from other social scientists and observers—corroborate the meager ethical moorings of the millennial and postmillennial generations.

It is difficult to determine whether the present situation is unique. Perhaps young persons in the United States have always suspended their ethical sense until they have reached positions of power or grown older and wiser in the pursuit thereof. Perhaps we are simply observing the swing of a pendulum that will soon revert to equilibrium. But, as far as I am concerned, these hedges are beside the point. The current situation of irresponsibility, of compromised work, is worrisome, whether or not it is unprecedented. I recall the words of former Federal Reserve Chairman Alan Greenspan: "It is not that humans have become any more greedy than in generations past. It is that the avenues to express greed had grown so enormously" (Goldstein, 2002). No doubt we are dealing with a human proclivity toward crossing ethical lines—the question is whether, and to what extent, that proclivity can be curbed, and an ethical sense can be cultivated instead.

Be that as it may be, our study of good work, which began as a program of social scientific research, raised questions of policy and practice. We wondered whether the present perilous situation could be ameliorated. Could we help to preserve good work in domains that seem well aligned? Could we provide support in domains that are poorly aligned? And, in particular, could we raise consciousness among young persons about the importance of carrying out good work and the perhaps less tangible but ultimately more satisfying rewards of doing so?

Here, in a way that we did not anticipate, the market proved relevant, perhaps all too relevant. We had a product: the results of our Good Work research, along with nascent ideas about how one might intervene to increase the incidence of good work. The question, baldly put, was this: If we issued a supply of good work findings and recommendations, would there be any purchasers, any buyers, of our product?

Briefly put: Our work has had virtually no resonance within genetics; the domain has remained relatively well aligned and many experts on genetics are already in place to monitor its progress along with any signs of turbulence. (Note, 2024: The situation may differ in the period ahead, because of controversies about which lines of work might better be suspended, because of unforeseen negative consequences.)

On the other hand, our work has been of considerable interest to journalists. In part, this appeal may be due to the fact that journalism is a troubled domain, we recognized it as such, and we had offered some ideas that might be of help. A major factor was a happy alliance with respected journalists Bill Kovach and Tom Rosenstiel, co-heads of the Committee of Concerned Journalists. William Damon worked with the Committee to develop a "Traveling Curriculum" (Bronk, 2010) that has already been used in approximately one-third of the print newsrooms in the country.

A word about the Traveling Curriculum: The major participants in the newsroom—reporters, editors, publishers—meet off-site for a day or two. In this deliberately unfamiliar setting, they discuss cases that pose ethical issues: for example, how to cover a story in which one has a personal interest; how to handle pressures from a lead advertiser; what to do when the Internet has scooped a lengthy and costly investigation; whether and how to cover a story that could be destructive for the community. While it is not easy to document the effects of such a workshop, participants report that the discussions are generative and that impact continues to reverberate months after the intervention.

What of the other domains that we have studied? To some extent, the story is marked by serendipity. Because we had considerable funds to study philanthropy, and because we have published about our findings in philanthropy (Damon & Verducci, 2006), there has been interest in this work—particularly in our chronicling of the "harms done in the name of philanthropy" (see Essay 31). Conversely, perhaps because we never received sufficient funding to study law and medicine in depth, and because, like genetics, these powerful fields are much studied, our modest work in these two "mega-professions" has received only sporadic attention. Thanks to the efforts of Joan Miller, a professor of nursing in Pennsylvania, an investigation was undertaken of good work in nursing in the United States and abroad.

Perhaps not surprisingly, since we are primarily educators, and since two of the principal investigators teach in schools of education, considerable interest in our work has arisen among educators. Our studies of higher education have engendered attention. As mentioned earlier, these studies have taken as a point of departure the institution—the college or university—rather than the individual worker. And much of the analysis has involved a comparison of the goals and values of the various constituencies on campus. Examination of the extent to which various stakeholders have similar views of the school, and consideration of what to do when these views do not mesh, has proved illuminating and suggestive. For recent work in this tradition, see Fischman and Gardner, 2022.

Paralleling efforts with the Traveling Curriculum in journalism, my longtime colleagues Lynn Barendsen and Wendy Fischman have developed an intervention called the GoodWork® Toolkit (pz.harvard.edu/resources/the-good-work -toolkit). The Toolkit consists of a set of nutritious, genuine cases organized in discrete chapters. The chapters map loosely onto the principal sections of the GoodWork interview: beliefs and values, goals, responsibilities, mentors and role models, the nature of excellence, and the overall contours of good work. Teachers or other staff who use the Toolkit are offered a wide set of options on how to proceed: they may, for example, go through the kit from beginning to end, select certain chapters for focus, add or revise cases, integrate cases into regular curricula when appropriate topics arise, or link the cases to events in the breaking news or on their own campus. (Note, 2024: In the succeeding decade and a half, a great deal of work has been done with the toolkit—see thegood-project.org.)

Let me convey a feeling for the Toolkit by summarizing three specimen cases, each of which actually happened:

- Debbie, the editor of a newspaper at a highly regarded independent school, learns about a rape on campus. She feels an obligation to report this incident. But the head of school pressures her not to report the rape, because it might dissuade those who are considering attending the school next year. Intensifying the conflict, Debbie comes from a family of intrepid journalists whose loyalty belongs to reporting the story, but she also has a younger brother at the school and does not want to jeopardize his standing.
- Allison is a high school scientist with a strong interest in neurobiology. Her heart is set on winning the Intel Talent search, with the concomitant scholarship for college. Her research involves an experiment with mice. Just before submitting her entry, Allison learns that the judges of the competition frown upon candidates who work with live animals. Determined not to jeopardize her chances for the scholarship, Allison redescribes her study as involving an examination of mouse behavior that she had viewed on videotape. To her delight, she wins the scholarship and proceeds to pursue a scientific career.

- Steve is an engineering professor at a liberal arts college. He is an excellent teacher and one who believes in honest feedback and rigorous grading. However, his tough grading practices result in students having less strong records when they apply to graduate school. Steve must decide whether to adhere to his demanding standards or whether to inflate grades so as not to penalize those students with professional aspirations in engineering.

Accompanying the cases are questions and provocations designed to sharpen the horns of the dilemmas and to stimulate the students to reflect on possible courses of actions and probable consequences thereof. For example, students are asked to role-play the various parties involved in the case of the high school rape and the reporting thereof. Or they are asked to outline the various options that Steve could follow if he decided not to alter his customary approach to grading. Or they are asked to predict what Allison will do the next time that she faces a moral dilemma, thus raising the possibility of a "slippery slope" culminating in chronic ethical malfeasance.

Students find the cases in the Toolkit engrossing. They like the cases, readily enter into the spirit of debate, often disagree quite vigorously with one another, and draw upon the cases and emerging concepts from one session to the next. A welcome and less predictable finding is that teachers also find the cases to be fascinating. Though prepared with a young audience in mind, the dilemmas resonate as well with older persons—and in fact, those cases stimulate them to reflect on their own career choices and on the dilemmas that continue to arise within their chosen profession.

You might think—and the creators might hope—that the Toolkit results in a rapid conversion to totally ethical courses of action. However, the early results prove far more complex, and, perhaps, far more intriguing.

To begin with, the cases direct attention to issues that turn out to be more vexed than might have been expected. By design, in most cases the best course of action is by no means clear, and especially not so for young persons who have had little experience in reflecting on issues of morality and ethics. Then, when young people do begin to talk about the issues, they discover that they may well not agree with one another—indeed, the disagreements may overwhelm the agreements, and cognitive paralysis may ensue.

A powerful additional force often enters the equation: the proclivity of adolescents—and, perhaps particularly, of male adolescents in the United States—to assume a relativistic or an antagonistic stance. Quite frequently the discussion begins with a bald statement that "that's the way the world is" or "ethics are a luxury we can't afford" or "who tells the truth, anyway?" Such sharply worded positions can produce a pointed rejoinder; but they can also serve to silence or mute an alternative, more nuanced point of view.

As a consequence, then, an early effect of the Toolkit may be to yield confusion rather than clarity, ambivalence rather than absolutism or absolution. Indeed, a study of a secondary school course on social justice, carried out by Scott Seider (2007), suggests that enrollment in such a course may actually make students *less* tolerant, less oriented toward social justice perspectives, at least in the short run.

There are many possible reasons for this paradoxical effect. I lean to the speculation that such a set of exercises exposes students to their own shortcomings, including their own self-centeredness or selfishness. Rather than face directly their own inadequacies, many students prefer to "blame the victim" and to assume a more judgmental and peremptory attitude rather than a more generous stance. As psychologist William James might have expressed it, "tough mindedness trumps tender heartedness."

Let me offer some observations, based on our opportunities to teach about Good Work at three different educational levels. For students at the secondary level, many of these issues are quite new. Sometimes they have not thought about them at all, and in such cases, even the introduction of a case may constitute a "treatment." Over the course of a session or two, the students' eyes are opened to considerations that they have not confronted before, with consequences that are not easy to predict. Note that we do not use words like "ethics" or "professions"; we just introduce the cases as scenarios that have actually happened and ask the students to make sense of them.

The ambience of the school itself becomes very important. School communities are generally intimate enough that a prevailing ethos—or lack of ethos—is apparent. When school leaders decide to devote efforts to promulgating issues of good work, this stance in and of itself is likely to have a significant effect on the student body. A decision, for example, to hold a workshop on "Why be honest?" or to devote an orientation at the start of the year to "Meaningful Work" constitutes a powerful statement.

The discussion about the Toolkit can also highlight fault lines in the community: If parents, students, teachers, and administrators are not of the same mind, this fact becomes apparent quickly. Indeed, members of our research team have quipped that we are most valued at places where we are probably not needed; in contrast, the places that could most benefit from discussions of Good Work may be too dysfunctional to take advantage of what we have to offer. The presence on campus of one or more champions of the project is essential.

In secondary schools in the United States with any degree of selectivity, one project of the students dominates all others: gaining admission to college. Indeed, for many students and even some families, this campaign is the most important project of their lives. The process need not be a damaging or hurtful one; students can learn about themselves, put their best foot forward, gain admission to the college of their dreams, and make the most of their opportunities. But all too often, the process is plagued by stress, dishonesty, and inequities: students distort their records or get excessive help in preparing the essay or managing the campus visit and the interview. And, alas, this dishonest enterprise becomes a model for carrying out work in the future—work that may end up being compromised if not frankly irresponsible.

Once in college, of course, there is the temptation to relax, and some students do just that. For others the process exemplified by college admissions simply persists, while being ratcheted up to the next level: Will I get into medical school, will I get the internship at a major law firm or newspaper, will I win the Rhodes Scholarship or the position with the management consultancy of my dreams? Good or poor habits acquired during the process of college admissions transfer all too readily to the next challenge at hand. (See the recently published findings in Fischman and Gardner, 2022, as well as extensive discussions at therealworldofcollege.com.)

Colleges and universities tend to be far larger than secondary schools, and students are older, far more mobile, far less accountable on a day-to-day basis: accordingly, the challenge of creating a community—and, in particular a "good work" community—proves more formidable. More likely, there is no pervasive ethos, but rather an informal collection of diverse and often shifting communities— an increasing proportion of them virtual, in this day and age.

The tertiary institutions that succeed in creating an integral community against the odds are ones that have long, self-conscious histories and deeply entrenched procedures and rituals into which students are acculturated, as well as regular processes of renewal. Princeton's preceptorial, the "Morehouse Man," and Swarthmore's hour of silence come to mind. Not all of these processes are necessarily benign: For every school that promotes community service or intellectual ambition, others catalyze or tolerate weekend binges or exclusionary access to fraternities or final clubs.

Some impressions: It is important that the Toolkit be introduced by faculty members who are trusted by students and skilled in leading discussions of multistrand issues, in a setting that is relaxed and that promotes reflection and camaraderie. Abstract discussions of issues like ethics (a word that we generally avoid) or responsibility are far less effective than dissection of concrete cases that engender deep and sometimes conflicting impulses. In addition to having a champion on campus, it is vital to create a feeling of belonging in the group— otherwise, these voluntary sessions tend to peter out. A focus on the work of college—grading, membership in clubs, treatment of diverse populations, finding a niche in the extracurricular terrain—is more alluring than a focus on work life after college. Still, issues that bridge these worlds—for example, how to select and benefit from a summer internship—compel interest on the part of most students.

It is important to acknowledge that many students carry out works of charity or community service while in secondary school or college. And while some of this work is doubtless résumé-packing, much of it seems genuine and meritorious.

Yet the excellent and often sacrificial work carried out by young Americans must be tempered by two considerations. First, students often couple a trust and helpfulness in regard to those within arm's length with cynicism or even outright hostility toward broader societal institutions, such as those of the government, the media, or the American corporate-sector life. Morality stops at the exit from the dorm, or, if one is lucky, at the border of the campus.

Second, and relatedly, students often execute good work without attention to what is happening in the broader society—they don't follow the news, they don't vote. Thus, as civic leader John Gardner once commented to our research group, "they may be helping dozens or even hundreds of people—and yet laws are being passed, or not passed, that are damaging thousands or even millions."

Finally, let me say a bit about teaching Good Work at the graduate level. For several years, I have taught the concepts of good work to my own students. Most of the students are educators and, in a sense, they have already committed themselves to a life of service—typically one in which they seek to help those who are less fortunate, and one for which they will receive relatively modest financial recompense in a very wealthy country.

In comparison with their younger collegiate counterparts, these students are far more aware, and far more critical, of the current market-dominated American society. For the most part they resent what they see as the intrusion of market considerations into the educational equation—choice, vouchers, merit pay, high-stakes testing. They are also painfully aware of how difficult it is to remain engaged in one's work when there is little public appreciation of stressful conditions at the workplace, and, all too often, little support from one's colleagues (one person's commitment to carry out good work may even be seen as threatening by colleagues who are not themselves carrying their own weight).

Those in education are dealing with ethical issues on a daily basis, and typically, they solve these on their own. The opportunity *to reflect with colleagues* about these issues is valuable—sometimes invaluable. More so than happens with other professional groups, considerations of empathy arise. The capacity to empathize with students (and their families) is of cardinal importance in a "caring profession" like education. Except at the higher end of the age spectrum, education is also a largely feminized profession. Many educators feel that they are not treated with as much seriousness or respect as would be the case if the profession featured a more balanced gender representation. In the terms of psychologist Carol Gilligan (1982), educators lean toward an ethics of care, rather than the ethics of justice that we found in professions like journalism and law.

As I've indicated, the Good Work Project is largely an American undertaking—we are researchers from the United States, working primarily with subjects who are American. It is therefore notable that the project has engendered considerable attention abroad—indeed, it might even be maintained that, like the theory of multiple intelligences, it has been more visible abroad than in the United States. (Note, 2024: This trend has definitely continued.)

This internationalization of the Project has proved a boon to our thinking and our practices. While some of the conditions that led to the Good Work Project may be United States–specific, the issues raised clearly resonate in many parts of the globe. At the same time, the issues will surely take different forms in different societies, and many of the most promising interventions and solutions will come from abroad. I have spent a considerable amount of time in Scandinavia as well as Reggio Emilia, an Italian community that I have whimsically declared an "honorary member" of Scandinavia. In addition to the attractions of food, climate, and camaraderie, I believe that the most impressive models of Good Work have evolved in these corners of Europe. (I might add that, due to rapid demographic changes in these regions brought about by immigration, issues of good citizenship have now come to the fore.)

As I have listened to international students and colleagues, I have come to understand that the North Americans look for solutions primarily from the individuals; East Asians from the family (and from entities, like the corporation, that are analogized to the family); and Europeans from the state. As was illustrated dramatically in the 20th century, different regions of the world spawn different counterforces to the market—religious, ideological, communal. And today, both continental Europe and Asia are far more likely to restrict the market (and the Internet) in various ways than is the case in the United States or Britain.

The components of good work play out differently around the planet. In some societies, such as those of newly developed states, ethical issues are most at risk; in other societies, such as those that are already quite comfortable, engagement is fragile; and, in a rapidly changing world, our conceptions of excellence are continually being renegotiated. Indeed, unless they are conversant with the new media and comfortable with rapid change, even the most talented young workers will soon become anachronistic.

Good work is most elusive when conditions are changing quickly, and when our senses of time and space are being altered by technology. It might be thought that the United States would have an advantage because of our leadership role in technology, and yet at present parts of Scandinavia and of East Asia are more digitally sophisticated than we are.

In closing, I would like to leave you with my best guess about the conditions that are likely to lead to good work, and how best to determine whether you are, or can become, a good worker.

Good work depends significantly on three kinds of support. The first is *vertical*. Crucial are the values and models set forth by those in positions of power and influence: initially your parents and older relatives; soon thereafter your teachers and coaches; and ultimately your own boss or employer, as well as the leaders of the broader society. Occasionally, an individual can be inspired by a *paragon*—someone not known personally but whose examples nonetheless provides guidance. The more these individuals model good work and expect to find it in the behaviors and attitudes of their charges, the more likely that you will become a good worker.

Next comes *horizontal* support: your friends and peers when you are growing up, your co-workers at the place of your employment. Needless to say, you don't have complete control over those age mates; your parents have a say about your friends, and the employers determine who will sit alongside you at your workplace. Nonetheless, within those broad constraints, you do have a choice on who you hang around with, learn from, or spurn. Especially in the United States, but increasingly in other parts of the world as well, horizontal support proves at least as important as vertical support, and increasingly so as one grows from childhood to adulthood.

Finally, there are periodic *booster shots*, of a positive or negative sort. Whether you are embarked on a good work pathway or not, there will be periodic wake-up calls: something bad that happens at the workplace or in your life, or, more happily, an example of true excellence, courage, or sacrifice for a cause more admirable than one's own self-interest. By definition, these booster shots constitute wake-up calls, but it is up to you whether and how you react to them. And those reactions can occur at the individual level; or, in the case of a company like *The New York Times*, at the time of an editorial crisis; or in a corporation like Johnson and Johnson, at the time of a drug scare.

My colleagues and I have been inspired by the writings of the economist Albert Hirschman (1970). This wise authority contends that everyone owes his or her organization a degree of *loyalty*. But at a certain point, if an inadvisable course is being followed, it is necessary to speak up, to "give *voice*." And finally, if one determines that one can no longer be effective, or no longer live with one-self, then the time has come to *exit*. In the Good Work Project, we particularly honor those individuals who attempt to change the course of their organization for the better; and if that fails, who launch a new organization that embodies the values of good work.

Turning to a final question, how can *you* determine whether you are a good worker, carrying out good work? We recommend the administration of 4 measures, each beginning with the letter M:

a. Mission—What is the mission that I am trying to carry out in my work? Is it consonant with the mission of my peers and, if not, what can I do about it?

b. Model—Who are the individuals I admire and why? Do I turn to them effectively when I am uncertain about what to do? What lessons can I learn from anti-mentors, or tormentors? And if I have no single mentor, can I cobble together a viable model—a so-called "fragment-or"?

c. Mirror Test (personal)—When I hold up the mirror to my own professional work, and peer into it clearly (without squinting or distorting), am I proud of what I see? If my mother were to read in the newspaper about all that I've done, would she be pleased or ashamed? And if I do not pass the mirror test, what am I going to do about it?

d. Mirror Test (professional)—When I hold up the mirror to my fellow professionals, am I proud or ashamed of what I see? If I am proud, am I doing my part? And if I am ashamed, what can I do to confirm the core values of the profession, and, if possible, to steer it into a healthier, more aligned condition? Should I seek to become a trustee of my profession?

Whether we are 10, 50, or 100 years of age, good work is never fully achieved, and never totally beyond reach. Indeed, it is better described as a continuing process of education—self education and education by others—than as a one-time or lifetime achievement. Learning about other good workers can be helpful, and the interventions that we've designed may also make a contribution. Ultimately, it is up to you whether you are seen as a good worker. Many individuals deemed professionals actually violate one or more of the three Es; and many workers in the least prestigious of occupations nonetheless exhibit excellent, engaged, and ethical work.

I'm often asked about what would be a sign that our project has been successful. One answer that I give is this: "When people all over ask about whether someone is a good worker, whether she or he is excellent, engaged, *and* ethical, and when the answer *truly matters*: *That* will be a sign that we've achieved the goal of our project." In this context I note with pleasure that a Philippine educator, Joy Abaquin, now gives public recognition to individuals who exemplify good work in different spheres, thereby combining my long-term interests in excellence and ethics.

Meanwhile, I end with the words of two wise persons. The novelist E. M. Forster memorably said, "Only connect." The goal of our work is to connect the three connotations of the word *good*—excellence, engagement, and ethics. And each morning, the writer and radio commentator Garrison Keillor ended his *Writer's Almanac* with a pithy phrase, and I'll conclude the lecture series by quoting his words: "Be well, do good work, and keep in touch."

## REFERENCES

Bronk, K. C. (2010). Education for good work. In H. Gardner (Ed.), *Good Work: Theory and practice*. Self-published.

Csikszentmihalyi, M., & Nakamura, J. (2007). Creativity and responsibility. In H. Gardner (Ed.), Responsibility at work: How leading professionals act (or don't act) responsibly. Jossey-Bass.

Damon, W. (2009). *The path to purpose: How young people find their calling in life*. Free Press.

Damon, W., & Verducci, S. (2006). *Taking philanthropy seriously*. Indiana University Press.

Gardner, H., Csikszentmihalyi, M., & Damon, W. (2001). *Good work: How excellence and ethics meet*. Basic Books.

Fischman, W., & Gardner, H. (2022). *The real world of college: What higher education is and what it can be*. MIT Press.

Gilligan, C. (1982). *In a different voice*. Harvard University Press.

Goldstein, B. (2002, July 21). Word for word/'Greenspan shrugged'; When greed was a virtue and regulation the enemy. *The New York Times*.

Hirschman, A. O. (1970). *Exit, voice, and loyalty*. Harvard University Press.

Seider, S. (2007). *Literature, justice, and resistance: Emerging adolescents from privileged groups in social action* [Doctoral dissertation, Harvard Graduate School of Education].

# THE PROFESSIONS

Within the Good Project (earlier the Good Work Project), my own interest has focused very much on the major professions in American life (see Essays 28–31).

But it's possible that work is not completely described as either good work or compromised work. In an essay co-authored with Laura Horn, I ponder an area of work that is extremely important in American society, but one that's rarely been studied carefully—philanthropy. Describing it as "the lonely profession," we analyze what it is like to work in area where standards are vague, and where one almost never gets honest feedback. We survey the reasons why individuals enter the field, what happens when they stay there, the sources of satisfaction, the sources of frustration, and "life after philanthropy."

Though our empirical investigation of good work in the professions ended early in the 21st century, I have been heavily involved in thinking about the professions in the succeeding decades—how they are impacted by the ubiquitous and powerful forces of the marketplace and, even more, by the advent of even more powerful AI programs. Not only do these computational "deep learning" capacities carry out much of the routine work in the professions, but they hold the promise of performing the more contemplative and integrating facets as well.

# THE PROFESSIONS

# Compromised Work

One would like to find an abundance of good workers across the professions: teachers who have mastered their subject matter, present it well, and behave in a civil manner toward students and peers; physicians who are knowledgeable about the latest techniques and medications and who cater to the ill no matter where they are encountered and whether they have resources; lawyers who can argue a case persuasively and who make their services available to those in need, irrespective of their ability to pay. Occasionally the impressive achievements of such individuals are publicly honored, and those concerned about the long-term welfare of the society hope that aspiring teachers, physicians, and lawyers will have ample exposure to such exemplars of good work.

Not surprisingly, the absence of good work commands the attention of scholars, journalists, dramatists, politicians, and ordinary folk. We are, perhaps naturally, perhaps understandably, fascinated to learn about the teacher who did not receive training in the subject that he teaches or who seduces a student; the physician who fakes her credentials or operates on the wrong patient; the lawyer who skirts the law or only defends the wealthy. As a friend quipped at the time, Time Warner might sell more copies if it renamed its venerable business publication *Misfortune*.

In the Good Work Project in which my colleagues and I are involved, we focus on those individuals and institutions that aspire toward, and in the happiest case exemplify, good work. Yet it is important to recognize that many individuals fail to achieve good work, that some do not even strive to be good workers, and that in the absence of compelling role models, future workers stand little chance of becoming good workers themselves. Hence, it is justifiable at times to suspend our focus on good work to see what can be learned from frankly deviant cases.

In what follows, I focus on what we have come to speak of as "compromised work." We conceptualize this variant as work that is not, strictly speaking, illegal, but whose quality compromises the ethical core of a profession. We do not concern ourselves with individuals who merit the descriptor "bad workers"— the journalist who steals, the physician who commits assault and battery, the lawyer who murders. Presumably these individuals would engage in such illegal acts irrespective of their professional status, and it is the job of law enforcement officials, and not of professional gatekeepers, to call these miscreants to account.

Rather, our concern is with the journalist who borrows stories, the politician whose word has no warrant, the physician who fails to heed the latest medical innovations and thus provides substandard treatment. Each of these individuals may at one time have embraced core values—journalistic integrity, political veracity, medical acumen—but at some point turned his back on the profession. If we can better understand how once-good workers begin to compromise their work, we may be able to enhance the ranks of good workers.

It is easiest to spot compromised work in professions that have existed for some time and whose principal values are widely shared. In such domains there should be consensual processes of training, recognized mentors, and established procedures in place for censuring or ostracizing those whose work violates norms of the domain, with disbarment or loss of license as the ultimate sanction. Of the three professions I will treat in this essay, law is closest to the prototype; journalism is furthest (many journalists lack formal training); and accounting (or auditing) is somewhere in between.

Since our project began (and no doubt long before), the pages of the newspapers have been filled with examples of compromised work; indeed, in preparing this essay I have sometimes been tempted to clip a healthy percentage of the stories in the daily newspaper. Here I focus on three cases from recent years that caught both my attention and that of the broader public.

The first case involves Jayson Blair, an ambitious reporter for *The New York Times* who was fired after it was discovered he had plagiarized and fabricated numerous stories. The second case centers on Hill & Barlow, a venerable Boston law firm that closed abruptly when its profitable real estate department announced it was leaving the firm. The third case centers on the flagship accounting firm Arthur Andersen, which went bankrupt after the Enron scandal of 2001.

In my initial study of compromised work, I chose these cases because they apparently represented three levels of analysis: Jayson Blair as an instance of compromised work by a single, flawed *individual*; Hill & Barlow as an instance of compromised work within a single *institution*; and the Arthur Andersen–Enron debacle as an instance of compromised work throughout a *profession*.

My study revealed, however, surprising continuities across these three apparently distinct levels of analysis. In each case, I found that I was studying individuals as well as institutions, and, indeed, an entire industry. Also to my surprise, I discovered that institutions held in high regard might prove especially vulnerable to the insidious virus of compromised work; I had expected that such institutions harbored "righting mechanisms" that for some reason had failed to detect the offending party. Finally, I expected that at least some instances of compromised work would be isolated and of relatively short duration. A far more complex and, to my mind, more troubling picture emerged—a picture that, moreover, reflects ominous trends in American society.

In 1999, Jayson Blair, a young African American with a flair for writing, became a regular reporter for *The New York Times*. Even before his stint at the *Times*, Blair had been regarded by peers and supervisors with a combination of admiration and suspicion. There was no question that Blair wrote well, had a nose for important stories, was a gifted schmoozer, and had impressed the governing powers at the college and community newspapers where he had previously worked.

At the same time, observers wondered whether he in fact had exercised the due diligence that is expected of a reporter; and indeed, supervisors had detected a highly unusual number of errors in his stories. While he had occasionally been admonished for carelessness, there had been few consequences. In fact, at the *Times,* Executive Editor Howell Raines and Managing Editor Gerald Boyd gave increasingly important assignments to Blair.

When Blair was discovered to have plagiarized a story from the *San Antonio Express-News*, he was immediately forced to resign. Then on May 11, 2003, in an unprecedented bout of self-examination, *The New York Times* devoted over *four full pages* to documentation of numerous cases of invention, plagiarism, and fraudulent expense and travel reports. Nor did the brouhaha over the Blair affair die down. Six weeks later, editors Raines and Boyd were themselves forced to resign their posts, and the new editorial regime at the *Times* explicitly dissociated itself from the policies and practices of its predecessors.

At first blush, Jayson Blair seemed to be an isolated case—a reporter who refused to play by the rules and who may well have suffered one or more illnesses. And in fact, there is ample evidence that Blair was a troubled young man who should have been carefully scrutinized for years. He was so unpopular at his college newspaper that he was relieved of his editorial position. When he was an intern at *The Boston Globe* in 1996–1997 and a freelancer there in 1998–1999, the sloppiness of his coverage was discussed. Shortly after he began to work full-time at the *Times*, Metropolitan Editor Jonathan Landman sent around a note that said, "We have got to stop Jayson from writing for the *Times*. Right now."

Blair soon accumulated a record number of corrections and complaints about his coverage. His behavior aroused dislike and suspicion among many of his contemporaries. But despite ample warning signs, Raines and Boyd took him under their wings; he was praised and offered ever-more-important assignments. And, to the shame of the *Times*, the decisive discovery of plagiarism was made not by its own staff but by a reporter for a regional paper.

To be sure, Blair had been a bad egg whose misbehaviors were more flagrant than those of his contemporaries. But at least since publisher Arthur Sulzberger Jr. had appointed Raines as managing editor in 2001, a strong set of explicit and implicit signals had been sent to the *Times* staff. Reporters were told they had to increase the "competitive metabolism" of the news coverage. Those who wrote flashy, trendy stories were rewarded with promotions, special privileges, and ample front-page coverage. In contrast, reporters who took a more thoughtful, less sensational approach, who emphasized the journalistic precept of carefulness, found themselves increasingly marginalized. Nor was this new culture a secret: In a much-discussed portrait of Raines that appeared in *The*

*New Yorker* in June 2002, the changing milieu at the *Times* was detailed and critiqued.

Had Jayson Blair been a truly isolated case, it is highly likely that the Sulzberger-Raines-Boyd managerial team would have survived intact and perhaps continued its questionably hectic pace and excessively dramatic bent. Once the Blair case broke, however, other heroes and casualties soon emerged. The most flagrant consequence was the abrupt resignation of star reporter Rick Bragg, who was accused of using unacknowledged stringers and of embellishing his lengthy and highly evocative stories. While Raines and Boyd fought to keep their positions, it was probably inevitable that sooner or later they would be squeezed out. The replacement appointment of Bill Keller—an individual widely considered a contrast in temperament and journalistic values—served as a sign that the *Times* was rejecting the "go-go" atmosphere of the previous few years.

Under Raines and Boyd, the *Times* had been engaged in an example of what I will call "superficial alignment." The editors were looking for young reporters who exemplified the pace and coverage they sought; the fact that Blair was African American was a bonus and, by the editors' own admission, caused them to cut him slack. For his part, Blair was keen at discerning what his editors desired; and, as befits an accomplished con man, he knew how to give the impression of good work and to cover his tracks. What both sides avoided in this *pas de deux* was a genuine alignment that honored the tried-and-true mission of journalism. Had Blair been subjected to a mentoring regime of tough love, he might have turned into a genuinely good reporter. And had he somehow slipped through an otherwise well-regulated training and supervision system, it is unlikely that the discovery of his misdeeds would have caused such turmoil in his company and, indeed, across the wider journalistic profession.

During the first week of December 2002, longtime residents of Boston were astonished to learn that the prestigious law firm Hill & Barlow had closed down the previous weekend. The firm had been in existence for over a century, was esteemed in the community, and had comprised in its legal ranks many prominent citizens, including at various times three governors of the Commonwealth. With their deep involvement in the community—exemplified by their defense in the famous Sacco-Vanzetti case of the 1920s—Hill & Barlow partners epitomized what legal scholar Anthony Kronman has called "lawyer statesmen." For outsiders, there was little reason to suspect any significant problems at Hill & Barlow—and none whatsoever to prepare them for its sudden dissolution.

A word about partnerships: Examination of over 1,200 hundred interviews in the eight domains considered by the Good Work Project reveals that *only lawyers* speak regularly about partnerships. In part a financial arrangement, in part a social network, the partnership serves as the locus for daily activity, the attraction and sharing of clients, and the mechanism for services and payment. The transition from associate to partner is the legal equivalent of the attainment of tenure in the academy; and in many ways, partners behave like members of a faculty. Young lawyers serve as associates until, assuming a good record and available slots, they are welcomed into the partnership, which is likely to be their home for the remainder of their professional lives. It goes without saying that the health and stability of the partnership is crucial for its constituent members, staff, and clients.

Each partnership has an institutional culture, passed on both explicitly and implicitly from the older partners to the new members of the association. By all reports, the institutional culture of the Hill & Barlow of old stressed intellectual and legal excellence; community service, including the holding of elected or appointed office; and a willingness to earn somewhat less money than competitors, in return for a lifestyle that was more balanced and that went beyond the sheer number and rate of billable hours.

Outsiders' initial reaction to the sudden closure of Hill & Barlow was a shock. After all, this was a partnership that had been highly esteemed for decades. To observers and the media, it appeared that overly avaricious lawyers from the real estate division had issued a fait accompli to their bewildered colleagues, thereby by one act destroying a distinguished New England law firm. The shock was compounded by the fact that the remaining partners did not even try to reconstitute the firm, but instead interpreted this mass exodus as a sign that the firm could no longer survive.

Closer examination reveals that the problems went back many years, perhaps several decades. Through the middle of the 20th century, Hill & Barlow did indeed have a deserved reputation as a firm of outstanding "lawyer statesmen" who not only were leaders in litigation and trusts, but who also stood out for their service to the community.

Yet, in my and my colleagues' analysis, this sterling reputation turns out to have been a mixed blessing. By the 1970s and 1980s, the situation in law had changed dramatically throughout the land. Whether lamented or not, the era of the lawyer statesman was over. Law firms were becoming much larger and more internationalized; corporate law divisions and the high-metabolism specialty of mergers and acquisitions were growing more rapidly than other spheres; many large corporations built up their own in-house legal teams; and individual lawyers were becoming far more mobile, as opportunities to make very large salaries materialized for those willing to jump ship.

None of these trends in itself necessitated a de-professionalization of the law. And indeed, many moderately sized law firms in New England and elsewhere took steps to modulate these trends: They increased in size or developed distinctive niches; they actively sought large corporate clients; and they reconfigured salary schedules to reward those lawyers who brought in the most business.

Perhaps most importantly, the more reflective firms realized that law was becoming more of a business: they recruited or trained professional managers; they were sensitive to the clout of specific partners and divisions; they paid close attention to changing patterns of income and expenses; they established governance vehicles whereby the most important members consulted regularly about trends and how best to meet them; they favored frequent, open, frank communications about all matters that materially affected the firm; and they were prepared, when necessary and with regret, to retire or marginalize partners who could not in any demonstrable way—for example, through attracting clients or managing the staff or training fledgling lawyers—contribute to the well-being of the firm.

According to our interviews with former members of Hill & Barlow, the firm did not seriously undertake any of these measures. Members continued to take pride in the history of the firm, and many continued to serve the community in various ways. But they did not work any longer as a firm of dedicated partners (epithets such as "a hotel for lawyers" and "university-style governance" were used by informants). Costs spiraled, but steps were not taken to increase income commensurately (or to lower costs, for example, by reducing the number of associates or moving to less luxurious quarters). Most damaging, the law firm never was able to create a governance structure that was widely respected by its members and that could meet these various challenges. On my analysis, it was the combination of the inordinately successful real estate group, on the one hand, and the ensemble of dysfunctional governance structures, on the other, that made the firm's closure inevitable.

I do not conclude that the Hill & Barlow partners necessarily compromised their practice of law per se. I do believe that both the real estate division, and the remaining partners who failed to deal decisively with the shifting terrain, undermined law as a profession. In acting in their own self-interest, they contributed to the destruction of the accumulated wisdom, public service emphasis, and pluralistic view of legal practice that had once characterized Hill & Barlow. To the extent that law simply becomes a collection of free-agent practitioners, for sale to the highest bidder, or a set of employees of multinational corporations, it will indeed be a diminished profession.

With the widespread use of double-entry bookkeeping and other financial and business innovations, the practice of accounting became a technical rather than a back-of-the-envelope practice in the 17th and 18th centuries. With the rise of corporations well over a century ago, and the advent of increasingly complex taxation and investment policies, the role of the independent certified auditor gained steadily in importance. Particularly at times of crisis, such as the stock market collapses at various points of the 20th century, the public was reminded of the importance of the accounting professions. Perhaps to his advantage, the auditor was seen as a rather colorless individual who followed technical rules in the manner of the archetypical Dickensian clerk or Weberian bureaucrat.

Within the profession and among those with close ties to the profession, there was keen awareness of crucial shifts that began in the 1970s. The wall that had once separated auditors from the firms they were monitoring had begun to crumble. Increasingly, personnel circulated between accounting firms and well-heeled client firms. Accounting firms set up consulting branches that worked with client firms; over time, the amount of consulting business often equaled or even surpassed that dedicated to the monitoring of the books.

In the go-go financial milieu of the 1980s and 1990s, markets became increasingly dominant in many spheres of life. Indeed, at the end of the 1990s, I made a quip that turned out to be uncannily prophetic: "If markets come to control everything, in the end there will be only one profession—accounting. And that is because only the auditors will be able to tell us whether the books are on the level or have been cooked."

But like most of the public, I was unprepared for the huge accounting scandals that captured the headlines at the start of the 21st century. Led by the renowned firm Arthur Andersen, the major firms were shown to have abandoned their professional disinterestedness (or "independence," as it is referred to in the profession) in flagrant ways. It was no longer unusual for accountants to hold stock in, work for, or consult for the firms they were allegedly monitoring; and for their part, firms went out of their way to provide lucrative work and extra perks for the supposedly independent auditors.

The smoking gun was the relationship between energy giant Enron and the flagship professional services firm of Arthur Andersen. These firms met powerful sanctions: bankruptcy with possible jail terms for those high-level managers whose involvement crossed the line from compromised to frankly bad work. Other major accounting firms also had to pay significant penalties; punitive new regulations and legislation were put into place; and many other business firms underwent probes or even dissolved. Meanwhile, the tacit or demonstrable complicity of members of boards of directors was amply documented, and the domain of accounting as a whole became very much under suspicion, its standing as a profession open to strong challenge.

The core value of the profession of public accounting is captured in the descriptor "public." Accountants receive training, licenses, and status commensurate thereto on the assumption that they will represent the public's interest in their review of the financial practices of individuals or corporations. Should the books appear questionable in any way, it is the duty of the public accountant to raise questions to the responsible individual or corporation and, if necessary, to refuse to certify that the accounts conform to generally accepted accounting principles.

Whether one thinks of journalism, law, or accounting, it is tempting to posit a golden age—a time when professionals were professionals, and the vast majority exemplified the highest values of the domain. But the mixed reputation of lawyers and journalists over the decades reveals the superficiality of such an analysis. And when one examines the history of accounting in the United States in the 20th century, one also discovers an oscillation between periods when auditors were under suspicion for questionable practices, and periods when corrective measures were installed and the prestige of the profession was restored. Indeed, such a swing of the pendulum can be seen in the history of Arthur Andersen.

As became well-known, Andersen had become the auditor for Enron. Widely touted as a model for a new kind of company for a new millennium, Enron trafficked in the selling of energy (especially gas) and energy futures. In 2000, it was, on paper, the seventh-largest firm in the United States, with a book value of $100 billion!

In 2001, the Enron bubble burst when it became clear that much of the corporation's alleged size, activity, and profitability were in fact fraudulent, the result of imaginative advertising and improper accounting. And when Arthur Andersen began to shred its Enron documents, the fate of the firm was sealed in the eyes of the media, the general public, and, eventually, the legal system.

Studies of the Andersen–Enron connection reveal that it had been deeply compromised for years. Enron was one of Andersen's largest clients; it paid a total of over $50 million a year to Andersen's auditing, consulting, and tax divisions. Employees shuttled back and forth between the two companies with such ease and frequency that it was sometimes difficult to tell for which they were working; at least 80 former Andersen auditors were working for Enron. The supposed line between the company being audited and the auditors evaluating the books of that company had become so blurred that, in effect, it no longer existed.

And yet it proved difficult to demonstrate sheer illegality. This is both because the nature of Enron's business was so new and so convoluted, and because so much of the role of the auditor/accountant remains an issue of professional judgment rather than of sheer legality or illegality.

In my view, the chief embodiment of compromised work in the accounting profession is the condition of *wearing two hats*—hats that inevitably pit key interests against one another. On the one hand, as representatives of the public, auditors and their umbrella organizations are supposed to remain at arm's length from the companies they monitor. On the other hand, the excitement and the monetary gains available for consulting prove irresistibly seductive for many auditors and their umbrella organizations. One cannot at the same time offer advice and feedback to companies while standing disinterestedly apart from their practices; in effect, one has become judge and litigant at the same time.

In each of the cases discussed, the background history covered a much longer period than I had anticipated. Jayson Blair's case reflected larger-scale trends at the *Times* dating back to the 1980s and exacerbated by the appointment of a new managerial regime in 2001; Hill & Barlow failed to recognize, let alone adapt to, forces that middle-sized law firms had been confronting for decades; and Arthur Andersen encountered longstanding tensions in the accounting profession regarding appropriate relations with clients.

Nor are the cases restricted to the particular examples on which I happened to focus: Within journalism, similar scandals had occurred in recent years at the *Boston Globe,* the *Washington Post, USA Today*, and the *New Republic.* Several dozen major law firms in Boston and elsewhere had either closed down or were absorbed into larger and more profitable firms. In recent years, each of the Big Five accounting firms saw significant scandals; comparable "multiple hats" problems arose in Europe and Asia; and compensatory legislation like the Sarbanes–Oxley Act caused turbulence in a great many American corporations. Whatever their usefulness for conceptualization and exposition, the three levels of analysis that I had selected turned out to be more closely related than I had expected.

If the study of "good work" is in its early adolescence, then the examination of "compromised work" is in its infancy. Firm conclusions would be decidedly premature. And yet, given the importance of the problem and its indissoluble links to issues of good work, a few summary comments are in order.

Because persons and institutions can go bad for any number of reasons, isolated cases of compromised work cannot be prevented. What is susceptible to treatment is *the soil in which compromised work is likely to arise and thrive.* Our three cases and others that could have been treated suggest that superficial signs of alignment can in fact be the enemies of good work. Respected institutions like *The New York Times,* Hill & Barlow, and Arthur Andersen create in their members—and in the general public—the belief that these institutions are inherently good and above suspicion. Those assigned the job of surveillance internally or externally may become lax, and, accordingly, those who are tempted to practice compromised work may find an unexpectedly promising breeding ground.

Indeed, these circumstances obtained in each of our three examples:

- Jayson Blair was on the make; Raines and Boyd wanted to remake the culture of the *Times* even at the cost of violating its most important values. And while various alarm bells tolled, none sounded loudly enough or insistently enough to be heard.
- Despite the enviable reputation of Hill & Barlow, many lawyers left the partnership starting in the 1980s; the particular requests of the real estate group were not taken seriously enough; and attempts to address the issue of financial survival and partnership communication were undertaken too late and with too little sense of urgency.
- Arthur Andersen had actually resisted temptations to enter the consulting world. But when it finally succumbed, it entered with a vengeance—and despite warnings about conflicts of interest. Spokespersons for the firm continued to enunciate the fundamentals of accounting, but too many partners and workers were trying to wear two incompatible hats. When the ambivalent Andersen encountered the swashbuckling Enron, a disaster was in the making.

In each case, superficial features and blandishments obscured the central values of the domain. During the Blair–Raines period at the *Times,* scrupulous and fair reporting was sacrificed to the immediately accessible and sexy. At Hill

& Barlow, the norms of an effective partnership were undermined as lawyers and entire departments went their own selfish way. And sometime in the last few decades, those responsible for the atmosphere of an accounting company forgot that it was supposed to be a public trust. Those on the inside should have seen these problems and made loud noises, but efforts to right the culture were too weak and ineffective. And so in each case it took a dramatic event—Blair's plagiarism, the real estate department's exodus, the Enron meltdown—to reveal what should have been clearer to those on the outside and clearest to those entrusted with preserving and embodying the values of the domain.

What happens when such a critical point is reached? It is possible, of course, that the domain will continue to deteriorate, and may come to be replaced altogether. Newspaper editor Harold Evans quipped, "The problem many organizations face is not to stay in business but to stay in journalism." The lawyer statesman no longer exists; it remains unclear whether he is being replaced by a viable option, or whether lawyers have just become high-priced free agents or cogs in a corporate legal machine. And if there are too many Enrons and the Big Five dwindle to Zero—it is not clear whether the books will be monitored in the future by independent accountants, government officials, private investigators, or powerful computational programs.

It is also possible that these professions will continue to survive but attract a different type of person with different kinds of values. With few exceptions, for example, broadcast television journalism exists as entertainment rather than as news. Totalitarian countries have bookkeepers, but, as the old joke goes, they produce "whatever numbers you would like us to produce." And it is certainly possible to have lawyer-whores who sell their services to the highest bidder. In such cases, those who want to know what is really happening in the world, whether the books are really accurate, or whether they can get a fair trial will no longer look to the members of the ascribed profession.

One goal of the Good Work Project is to help bring about a happier scenario. Professions will always feel pressures of one type or another, and, at the time of powerful market forces, these pressures can be decisive. The forces cannot be ignored; they must be dealt with—but they must not be succumbed to. Those individuals, institutions, and professions that actively cope with these forces while adhering to the central and irreplaceable values of the domain are most likely to survive and to thrive.

How to do this?

In our project, we speak of the four Ms that help to propagate good work The Ms seek answers to the following questions:

- What is the *mission* of our domain?
- What are the positive and negative *models* that we must keep in mind?
- When we look into the *mirror* as *individual professionals*, are we proud or embarrassed by what we see?

- When we hold up the *mirror* to our *profession*—or, indeed, our *society*—as a whole, are we proud or embarrassed by what we see? And, if the latter, what are we prepared to do about it?

I suggest that if the individuals and institutions described here had perennially posed these questions and tried to answer them in a serious, transparent way, they would not have become targets for our study.

# The Lonely Profession

*With Laura Horn*

Grantmaking seems like the ideal job! You earn a living by distributing large sums of other people's money. You work in a plush office with resources to fly around the world collecting the information you need to carry out your responsibility. People hang on your every word, laugh at your jokes, and treat you like royalty because they want the funds that you control. You don't have to raise money or worry about the bottom line, and, barring gross malfeasance, there is little risk of losing your job. You work on difficult social, political, or scholarly problems from a place of relative luxury and can feel good about yourself at the end of the day for your efforts to change society. Framed in this way, the job of a grantmaker sounds idyllic.

In truth, the life of the professional grantmaker is not as serene as it sounds. We've concluded that the current environment in organized philanthropy makes it difficult to find lasting satisfaction from one's philanthropic work. Professional grantmakers support other people's work by giving away other people's money and cannot legitimately take credit for the work that they fund or for the generosity of the philanthropic donor. Many grantmakers feel isolated from the public, their grantees, and their professional colleagues. The structure of philanthropy as a whole provides little grounding for its professional grantmakers.

Perhaps as a result of these conditions, most program officers and foundation executives do not approach philanthropy as a career. They do not plan to work in philanthropy, and once they are there, they do not plan to stay forever. Rather than allowing philanthropy alone to define their professional lives, the grantmakers and executives we interviewed take three different stances toward the field:

- They view philanthropy as a continuation of their established career;
- Or as an opportunity to pursue a specific agenda of personal importance;
- Or as a way of taking a broad look at the world.

Philanthropy does not define their work, but enhances it.

This characterization of a single calling cries out for a comparative perspective. Typically, professionals in other domains feel called to enter their respective

248

fields and work hard to get there. They go through a period of training to develop the competencies required to be accepted as members of the profession. In contrast to grantmakers, who may maintain a sense of professional identity independent of philanthropy, doctors, lawyers, journalists, and scientists, typically view their chosen professions as a lifelong career, feel passionate about their work, and—under normal circumstances—remain committed to carrying out the mission and upholding the standards of the profession throughout their working lives. While each profession inevitably presents obstacles that professionals must overcome to do work that they feel good about, practitioners normally stay rooted in the ideals of their professions, even in the face of these challenges (see Essays 28–29).

In what follows, we take a detailed look at a cross-section of those involved with philanthropy: program officers and executives employed by medium to large private foundations. Throughout this essay, the term "grantmaker" refers specifically to this group of philanthropic professionals. Our analysis draws primarily from interviews with leading figures in traditional organized philanthropy.

## EVIDENCE OF PROFESSIONAL IDENTITY

In comparison with those in other professions, we noticed distinct features in the way grantmakers talk about themselves in relation to the field as a whole. In stark contrast to the doctors, lawyers, journalists, scientists, and actors, for whom entering their professional fields often represents the fulfillment of a dream and a source of great pride, grantmaking is typically an unplanned occupational shift, rather than a long-contemplated career decision.

For many professionals, the formation of a professional identity begins when they choose to pursue a specific profession, well before they become members of their chosen profession. Many describe their professional work as a calling.

In contrast, grantmakers and foundation executives do not plan to go into philanthropy, as do their counterparts in other professions. They describe "backing into" the field, or ending up in philanthropy by "accident."

A sample characterization:

> People who go into philanthropy not only weren't prepared for it, but weren't even thinking about it. They were running their deanship or their department chairmanship or doing something, running a little nonprofit, and some foundation taps them on the shoulder and says, "We want you to come in and join us."

None of the program officers or foundation executives we interviewed had planned a career in philanthropy; rather, the opportunity to enter the field presented itself unexpectedly as they were pursuing other work. A well-known foundation executive describes his entry into the field of philanthropy:

I stumbled into this. I only had the vaguest understanding that there was even such a profession as philanthropy when I was going to graduate school. Of course I'd heard of the Ford Foundation and a few other foundations, but I had only a dim understanding of what in fact they did . . . It's the sort of thing when you do a lot of other things and then an opportunity arises and you can go into it.

A few of the grantmakers we interviewed frame their unexpected encounter with philanthropy as a lucky turn of events.

What I find interesting is the idealistic sense of trying to make the world a better place. And I never dreamed I would be doing it from [name of foundation]. I always thought I would be doing it as an activist in the community. And then I just got terribly lucky and ended up here.

More commonly, however, grantmakers accept positions in philanthropy with some reluctance. Consider this testimony:

I wasn't sure about coming to the [name of foundation]. I had not dealt much with foundations. I didn't have a very high opinion of foundations. I mean, in the issues I cared about . . . they were not necessarily working in some of the concerns that I had . . . I didn't necessarily want to be here, it was not my ideal to come here.

Or this characterization:

Well, I was a really hard sell, as the story goes around here, because I turned them down a couple times. I had always wanted to be a professor . . . I had an endowed professorship with tenure and a young family . . . and that's what I wanted to retire doing.

## PROFESSIONAL STANCE

Most grantmakers do not identify with philanthropy as a career in itself. They approach philanthropy on their own terms, with an independent sense of how philanthropy fits into their identity as professionals. In fact, we found three common stances that grantmakers take toward the field: They view philanthropy as a continuation of their established career; as an opportunity to pursue a specific agenda of personal importance; or as an opportunity to look broadly at the world.

### An Established Career

Some grantmakers have a career identity in a different field before entering philanthropy and regard philanthropy as a variation on their previous work rather than a different profession altogether. These individuals view their grantmaking work as one phase of their career in a different field such as academia, law, education, or public policy.

Some of these grantmakers work in philanthropy as a hiatus from their primary career. They either work in philanthropy temporarily before moving back to the field in which they were trained, or they work in philanthropy at the end of their career as a way to influence their field more broadly before retiring. Consider one subject who spent most of her professional life at a liberal arts college as a professor and administrator before accepting a grantmaking position at a foundation.

> I think that philanthropy in general probably wouldn't have attracted me. It was what this foundation did . . . I've been trying to do good within the liberal arts college sector for all of my life because I deeply believe in it as the very best way to educate the young. And I just see this as an extension of that work. It's just that I'm able to do it more broadly.

Others in this group work in philanthropy indefinitely, but remain strongly grounded in their previous career.

> There is, I think, a strong connection between what I did before and what I do now. I was an educator for 25 years before I started doing foundation work, and for the first dozen years here, in addition to being the president of the foundation, I was also the higher education program officer . . . I don't do that anymore, but that was obviously a very nice thread from my career in education through my work in philanthropy.

Later in the interview, he said:

> I think it helps to have had another career before you do this work so that this does not become your identity . . . [I have] a pretty healthy sense of self that if I couldn't do this tomorrow, that would be fine. I'd go do something else. I don't need [name of foundation] to define who I am.

## A Specific Agenda

Other grantmakers dedicate their professional lives to pursuing a personal mission and approach their work in philanthropy as one of many ways to advance that mission. They come to philanthropy with a specific agenda such as "encouraging youth involvement in communities" and act as if they could leave philanthropy at any time to pursue their agenda elsewhere.

A grantmaker illustrates this type of professional stance in the following passage:

> My work is an expression of what I believe and what I value. That's what guides me. In other words, like I said earlier, if I weren't doing it here, I wouldn't have the [name

of foundation] money to do it, but I would find another way to do some version of what I'm doing. It's a privilege to have this opportunity, but it's my values that are guiding that . . . So wherever I wanted to work, I wanted to do something for those who were . . . dispossessed, if you will, who have been left out.

For these grantmakers, philanthropy serves as an opportunity to enact values or to carry out goals that are not directly related to philanthropy.

The way the [name of foundation] works is that they bring in program officers to help shape the priorities, not just to carry out some already shaped priorities. So I had a chance to emphasize things that I thought were important . . . And the whole notion of an experimentalist world of solving problems, of being eternally optimistic, of being committed to democratic processes, that's who I think I am. And this was an opportunity. Being in the foundation was an opportunity to put those values into programmatic form.

## A Broad Perspective

In a third approach to philanthropic work, grantmakers enter philanthropy because it allows them to apply their capacity to think broadly and to pursue wide-ranging interests. They see themselves as generalists, and philanthropy allows them to maintain a broad perspective on the world. Some have been trained in a specific discipline and want to widen their focus. They use philanthropy as a way to step back from their narrow discipline, to "see across the fray" and take a broad look at the world before moving on to other work.

I had intended to become an academic. And then for one reason or another, I decided I didn't want to be an academic. I wasn't all that interested after a certain age in disciplinary research, which was—I felt that was what you had to do. Now, I can see now that you can make your way [in academia] without necessarily doing narrow disciplinary research, but I didn't understand that at the time . . . So I stumbled into this.

Philanthropy allowed this subject to look at the world through a wide lens to gain a perspective that he couldn't see during his previous, narrowly focused work. He plans to return to academic work, writing and teaching, once his tenure at the foundation expires.

Another subject described a similar approach to the field of philanthropy:

Before the war, World War II, I had been an intensely focused faculty member, psychologist. My world was psychology—my research, my students—I wasn't even interested in faculty meetings . . . but when I came back, I was determined to find work that exposed me to a broader range of social issues, social problems, the way the world functioned. And it was just extraordinary luck that [name of foundation] was holding a spot which they did not intend to fill until they interviewed me . . .

Others who approach philanthropy with this generalist stance have a broad education. They highlight their ability to see across different realms of life and to synthesize complex information. Philanthropy is a way to exercise these skills without narrowing their professional focus.

I have a PhD in American Political Thought, which equips me to do little else than to make judgments about broad and general ideas of political thought. And this is a way for me to exercise that particular training—that particular talent.

We have seen that most grantmakers prefer to think of themselves as professionals or workers with other backgrounds and interests who happen to be carrying out a stint in the world of philanthropy. This distance from their current profession may explain the feelings of alienation that they often expressed.

## WHY DON'T THEY IDENTIFY WITH PHILANTHROPY?

We have identified several factors that mitigate professional identification with the field of philanthropy. Grantmakers own neither the money they give away nor the work that they support with that money. They feel isolated from grantees, the public, and their colleagues. Foundation policies often limit the amount of time they can stay in the field of philanthropy, and the field as a whole lacks a unified set of professional values to anchor and guide them while they are there.

### It's Not My Money

Traditionally, philanthropy is the domain of the wealthy. New philanthropic models, such as donor-advised funds and giving circles, have made organized philanthropy more widely accessible; but still, most people making decisions about where to donate philanthropic money are connected to the money itself. In capitalist America, philanthropy is built on the fundamental assumption that donors have a right to give away money as they see fit, just as they have a right to accumulate as much wealth as they can. Philanthropic freedom is widely celebrated:

I think one of the greatest things about philanthropy in the U.S. is the freedom donors have to pursue their own vision, their ideas about how to use their luck and skill and what benefit it brought to them personally, to help support the hopes and dreams of other people who are still struggling.

Wealthy individuals are free to support public institutions of personal importance. Their philanthropic decisions are respected and rewarded financially through tax benefits. They derive their philanthropic authority from their generosity.

Professional grantmakers break from this philanthropic tradition. They give away other people's money. The right to give it away, however, does not apply as it would were they the donors. They must derive their grantmaking authority from something other than wealth. The legitimacy of professionals in any realm lies in the answer to the following question: Why should society reward them for what they do? Philanthropic donors are valued for their generosity, but why should society reward grantmakers with money and respect for giving away someone else's money? Philanthropic tradition has not established a clear answer to this question.

Again, the contrast with other domains proves instructive. The traditions of law and medicine provide lawyers and doctors with confidence that their work is of social value, that simply by carrying out the work they were trained to do, they legitimately earn the social rewards of money and prestige.

Because philanthropic tradition in America is based on the generosity of donors, not on the work of professionals, there is no such confidence among grantmakers, no collective sense that simply carrying out their daily work is enough to earn a respected position in professional society. By centering their identities elsewhere, they reassure themselves of their social value as professionals.

Within the field of philanthropy, grantmakers generally have less power than the donors and board members. Almost all of the grantmakers we interviewed emphasized that they are not giving away their own money. This recognition was not a simple statement of fact, but a salient aspect of their work that brings a complicated mix of emotional and interpersonal dynamics. They feel privileged to have the opportunity to spend money that isn't theirs on projects and ideas they deem important.

However, the will of the donor and the opinions of board members loom as potential limits on their professional autonomy. In most cases, grantmakers choose to work at foundations that are sympathetic to the specific ideas and causes they want to support, foundations that also offer them the autonomy to make grants they think will make the most difference. However, when their opinions on grantmaking diverge from those more closely connected to the source of the philanthropic money, the whim of a donor or board member can trump the grantmaker's professional expertise. The observations of one grantmaker reflect this tension between wealth and autonomy:

This is an enormously luxurious circumstance, full of supports. There are financial resources. There is autonomy. There is no ballot box that will vote us out of office. There are no shareholders that will vote us out of office at the next shareholder meeting . . . So the supports are enormous. Just as we are luxurious with all of those assets, the reality comes to play when you are obviously dependent upon a board to support and acknowledge and ultimately approve the work of the foundation staff.

Another grantmaker framed the involvement of foundation board members in grantmaking as a challenge in her work:

> Most foundations in this country still have boards that are actively engaged on the transactional side of grantmaking. And that can be an obstacle.

Adopting a professional orientation rooted in something other than the field of philanthropy gives professional grantmakers a greater sense of legitimacy in their grantmaking decisions. External experience and expertise are a grantmaker's key to professional autonomy.

Dealing with large amounts of wealth can be uncomfortable for grantmakers. Often the grantmakers' progressive political leanings run counter to the capitalist structure that created the foundation's wealth. One subject forcefully highlighted this dynamic:

> Another thing that may be very complicated and complicating is the relationship between the donors and the rich people and the staff of the foundation. If you don't work that out effectively, then it's very hard to make effective grants . . . Let me just be clear that I think one thing that limits the horizons for foundations is the fact that . . . almost 100% of them have rich people in decision-making positions . . . I think first of all what helps you make a bundle is that you really want to. I think people rarely make a bundle without that being a very core value for them. They get rich because they want to get rich. And they're working in a community where that value is common, widely shared. And it's unquestioned, largely unquestioned. So they're now going to a world where virtually everybody has said, "we don't want to make a bundle." We're like a professor, or community organization, or museum director. They're all people who've said, "we don't want to make a bundle." We have some sense of sufficiency, not maximization. Which makes us kind of strangers in this world.

Arrogance is commonly cited as one of the dark sides of philanthropy; in order to avoid the conceit associated with wealth, professional grantmakers distance themselves from philanthropy and root themselves in the fields they support. They would often prefer to be associated with the work of the grantees rather than with the wealth of the donors. Many subjects emphasized the distinction between grantmakers and philanthropic donors, and nearly all warned against identifying too closely with the philanthropic money.

Grantmakers' ambivalence toward wealth is one factor that can dissuade them from centering their professional identity in philanthropy. Grantmakers want to avoid the arrogance stereotypically associated with wealth, but at the same time, their distance from the philanthropic money they manage limits their professional autonomy in philanthropy. One subject summed up the connection between grantmakers' ambivalence toward wealth and their lack of identification with philanthropy:

> It's both exhilarating and uncomfortable to have access to this kind of money. I think every grantmaker I know has always felt like this is an incredible privilege. What they do with the emotions that they have tied to that is very different. I think some people can get extremely anxious and other people can get arrogant. Some people kind of despise the source. One of the things I think is really interesting about grantmaking is that it's something that people really love to hate. The money often was acquired in not very honorable ways. It is very much a capitalist tool and I think there is a lot of ambivalence about it. People have ambivalence about wealth anyway.

> As a result, I think a lot of grantmakers, their orientation is to their grantees and very rarely do they feel an orientation to the field of philanthropy or to their organizations . . . I think part of it is this ambivalence about not wanting to be seen a) as a philanthropist or b) as being too connected to just the raw money. You want to be connected to the outcome from having this money.

In order to gain professional legitimacy and to avoid the arrogance associated with wealth, grantmakers tend to shy away from defining their career as their work in philanthropy.

## It's Not My Project

Not only are grantmakers removed from the philanthropic money they manage, they are also one step removed from the social change they support with their grants. While doctors and lawyers can see the results of their work in a cured patient or a guilty verdict, grantmakers are inevitably removed from the fruits of their labor. The grantmakers we interviewed acknowledge that they are not directly responsible for the success of the projects that they fund.

> Well, I don't think you can develop a lot of pride out of your philanthropic activities in a big foundation. You are given this extraordinary opportunity to spend money that isn't your money, and it's just, there isn't anything you can congratulate yourself on except spotting good people.

> Another grantmaker presented a similar argument:

> The good foundation recognizes the fact that we operate not center stage, but off-stage. The good foundation recognizes that standing alone—it really accomplishes nothing. I mean when you really think about it. That it facilitates, that it abets, that it brokers, that it encourages, that it supports, that it celebrates the work of others . . .

You ought not lose sight of the fact that you don't play Brahms, you don't heal the sick, you don't educate the first kid, you don't do anything, okay?

Denying themselves credit for the work of grantees may be in part an exaggerated modesty, motivated by a desire to appear humble, to avoid the arrogance that can taint philanthropic work. Indeed, consciously cultivating humility among grantmakers is a wise strategy for good grantmaking. However, their humility also reflects a real, underlying challenge for grantmakers—how to claim some sense of professional accomplishment and competence without falsely taking credit for the generosity of some or for the work of others. All of our subjects seem to struggle with this dilemma, highlighted clearly by one subject in the following passage:

> I had to learn how to be comfortable being very far behind the scenes, living vicariously through the work of others . . . I hadn't done it before. I had never been three steps away from the action. And I think it took some time for me to learn. First I was comfortable with it, but then I had to learn how to do it. How do you stay connected to the real work? How do you stay rooted enough in the challenges of that work so you understand it? . . . There's an ongoing need for kind of a reality check about what you can and can't do from philanthropy. Again, part of the potential flaw of arrogance in philanthropy is that you hear from people that work in philanthropy when they say, "We're doing this and we're doing that." We don't do anything. We invest in people who do things.

## LONELY WORK

According to many of our subjects, grantmaking is a lonely endeavor. Associating with philanthropy can isolate grantmaking professionals from the very people they are trying to serve: the grantees and the public. Many grantmakers also feel alienated from their professional colleagues, both within the foundation and in the broader field.

### Isolation From Grantees

The power difference between funders and grantees puts grantmakers in the uncomfortable position of having to evaluate the sincerity of each interaction.

> It's very hard to establish a truly honest relationship in this position and that's very frustrating . . . I think this is a rather lonely profession because nobody's ever honest with you . . . somebody always wants something from you.

Trusting everything that people say about you as a grantmaker can lead to an overinflated ego. On the other hand, approaching each interaction with skepticism is a lonely way to relate to people with whom you are trying to develop a trusting collaboration, especially when you have to discount any praise that you might receive. Identifying with philanthropy makes the power differential more salient. The temptation to try to sidestep this isolating dynamic by identifying with something other than philanthropy is understandable. It may also contribute to grantmakers' reluctance to embrace philanthropy as a professional home.

## Isolation From the Public

Not only do grantmakers feel isolated from grantees and potential grantees, but they also feel remote from the public at large. The public perception that grantmaking is fun and easy does not match the working reality that grantmakers encounter as they try to navigate the field of philanthropy in pursuit of often-elusive social change. Foundation work can be frustrating for grantmakers with high personal standards, since the success of their work is difficult to measure.

As one subject put it,

> This is half in jest, but there's some seriousness to this, when I started to come work here occasionally people said, "Oh, it must be great working at a foundation, just giving away money, that must be lots of fun." And the fact of the matter is it's really hard giving away money and doing it well. You know, so this notion that you're just sitting there in a highfalutin' place giving away money—but that's people's notion of what foundations do. It is so far from the reality of what it is to really try and do this well.

Some grantmakers and executives also cited as a challenge a growing public suspicion about philanthropy. For people who enter the field of philanthropy with the best intentions to help society, public skepticism is a difficult thing to swallow; it may deter grantmakers from proclaiming their allegiance to the field of philanthropy.

## Isolation Within Foundations

Grantmakers often feel disconnected from their foundation colleagues. Most large foundations divide their grantmaking into content areas, or fields that they support, such as education, the environment, and the arts. Foundations often hire grantmakers because of their experience and expertise in other fields. They place emphasis on the areas they support and expect grantmakers to do the same. This fragmentation of philanthropy can isolate grantmakers from one another.

One subject from a foundation that recently shifted from one of general grantmaking to one organized by program areas noticed the type of unanticipated consequence that can arise from the emphasis on specialization.

Initially we didn't have program areas that were basically driven by individual staff people. We had a cadre of generalists who were working across the full repertoire . . . As we tried to become more strategic and become more proactive in terms of becoming more focused and more tactical in terms of our work, we started to hire people who were from the field. We said, "All right, you've done that wonderful work now in X University. Now come and do it in a foundation." And by definition we began to create these areas of specialty . . . And all the incentives were aligned in such a fashion that the wonderful expert was rewarded and supported for doing her work in that field . . . How could we incentivize cross-program work? How could we climb up out of our silos, embracing one another and saying, "Let's connect and do our work together."

Separating grantmakers by content area encourages them to expand their knowledge of other fields rather than connecting with their professional grant-making peers. This policy can have the unwanted effect of isolating grantmakers from one another, even within a given foundation, especially if the foundation's overall mission is vague, a challenge highlighted by several grantmakers. One subject felt an important shift in her work when the foundation clarified its overall mission.

I think the important issue that has changed is this: there's a clear sense of mission in the Foundation . . . The difference in terms of the hierarchy is that we all have to work in concert toward a mission versus before it was toward a program . . . you don't feel as hierarchical when your work is part of a bigger piece and parcel . . . So [previously] it was very much whatever an officer or a director wanted to do. Now there's an expectation that says, "How does what you want to do get us closer to this bigger, broader mission?"

## Isolation Between Foundations

Just as grantmakers are isolated by subject area within foundations, so they are isolated in the broader field. Collaborative efforts between foundations usually reflect the organization of foundations by grantmaking content areas. While collaborations by funding niche or geographical region are helpful, they rarely focus on best practices for the entire field, preventing a sense of camaraderie and field-wide support. As one subject stated, "Foundations are fiercely independent."

Grantmakers rely primarily on foundation culture for professional guidance; this provides only precarious professional grounding in philanthropy, since foundation culture can change dramatically with a simple change in leadership. If the alignment between a grantmaker's goals and the foundation's grantmaking philosophy breaks down, there is little fieldwide community to turn to for support.

Grantmakers and executives undoubtedly enter philanthropy with the best intentions to help society, to conduct work that is excellent in quality, socially responsible, and personally rewarding, but they are offered little support or guidance on how to channel their efforts effectively, wasting precious human and monetary capital.

The current structure of the field takes a toll on grantmaking professionals, and on the quality of their work.

In contrast to other professions, professional philanthropy lacks shared norms regarding the purpose and practice of grantmaking. Most grantmakers agree only on "positive social change" as the purpose of philanthropy, a concept so vague it lacks the power of a professional mission. As one grantmaker put it, "You could drive Mack trucks through any of these wonderful philanthropic ideals." More precise philanthropic missions cited by the grantmakers we interviewed, such as supporting grassroots social movements, creating new institutions, or building new fields of knowledge, are idiosyncratic.

Grantmakers also lack shared professional standards. What some view as effective grantmaking, others dismiss as bad practice. They disagree on the proper role of board members in grantmaking decisions, the appropriate involvement of program officers with the projects they fund, and the type of experience and training that adequately qualifies an individual to become a grantmaker.

While professional grantmakers often praise such fragmentation as pluralism and diversity, as the result of philanthropic freedom, the same individuals often long for more clarity in their roles and more fieldwide collaboration. A survey of employees at a large, well-known foundation highlighted the dilemma of many grantmakers:

> One of the discoveries that we made when we were doing a study of the Foundation to understand how people were balancing their work and personal lives was that a lot of people said, "Well, one of the things that's most difficult for us is the stress of not knowing what the job is."

While some grantmakers manage to find lasting satisfaction in their work with the support of good mentorship, personal reflection, enduring stamina, and unusual patience, many other grantmakers eventually burn out. As a result, many potentially good practitioners leave philanthropy, or worse, they stay and let their work suffer.

People do their best work when they enjoy what they do. And they are most likely to experience deep satisfaction and genuine enjoyment in their work—an experience described by Csikszentmihalyi as "flow"—when a job provides clear goals, immediate feedback, and a level of challenge that matches their skills. Along with excellence and ethics, the opportunity for flow experiences—for engagement—appears to be constitute a third element of "good work."

Professional philanthropy provides surprisingly little opportunity for flow, for engagement. Grantmakers are generally unclear about their roles, and even the most hardworking and well-intentioned grantmakers may not immediately see the direct results of their labor. When a domain limits opportunity for rewarding and enjoyable work experiences, the risk is that members of the field will become bored and retreat into a rigid orthodoxy in an attempt to protect the relevance of their work. Professional grantmakers must find a way to stay engaged with their work if they are to fulfill their aspirations to be effective and responsible grantmakers.

Some foundations anticipate the problem of burnout and limit the amount of time a person can work at that foundation, with the aim of getting people out of the field before they become too complacent or discouraged and do too much social damage. These policies reinforce the mindset that employment in philanthropy is a temporary endeavor that serves as a complement to one's professional activities outside of philanthropy, not a career to be pursued in itself. Inadvertently, this policy may exacerbate the defensive stance that professionals take toward the field.

Because grantmakers consistently encounter systemic obstacles to carrying out good work in philanthropy, it is no wonder that many grantmakers keep the field at arm's length, that they look for professional grounding elsewhere. But does it have to be this way? Could philanthropy be a place to look for professional identity, or is it destined to be a field for amateurs where passion reigns and people leave when they burn out?

The field of philanthropy is divided on this question. Neither grantmakers nor foundations want to give up the freedom that the lack of professional structure permits, but at the same time they also want more clarity in their own work, and more effective and responsible work from the field as a whole.

Perhaps there is a middle ground. Pluralism in philanthropy does not require isolationism. It is clear that actors in the field of philanthropy—professional grantmakers, board members, donors, and grantees—need to reach beyond their regional, philosophical, and content-based niches. Possibly such discussions, conducted seriously over a period of time, might yield real and perhaps even surprising consensus.

Relying exclusively on exceptional individuals to find their own way in professional grantmaking is not a sustainable strategy for good work in philanthropy. However, there is no need to have a single model of a philanthropic career. Reaching agreement on a few effective trajectories would provide enough guidance to ground professionals in their philanthropic work and provide them with the framework needed to carry out their best work, while still respecting the value of various philanthropic models.

# In Defense of Disinterestedness in the Digital Era

*As a student of the professions, I have long been convinced of the importance of maintaining a disinterested stance—not to take the easiest way, nor the quickest way, out of a dilemma, but rather to confirm the basic assumptions of one's profession and to seek to maintain them in the face of powerful and often contradictory pressures.*

*Essay 32 was written in the first half of the second decade of the 21st century. At that time, some of my colleagues wondered about the need to be distinterested—maybe one should just push what one believes in, whether or not it is evidence-based or ethical. Let the loudest voice prevail!*

*But in the succeeding years—which one might think of as "the Trump Era"—the importance of a disinterested stance was recognized as more important than ever, even as it was proving more difficult than ever to achieve.*

Imagine a world in which a physician routinely recommended medicines produced by a drug company that was supporting his research; a middle school teacher devoted the bulk of her time to a student whose parents had endowed (or had promised to endow) a building at the school; an auditor gave an unwarranted high rating to a company in which he had recently made an investment; a judge failed to disclose her relationship to one of the parties in a suit that she was hearing; a journalist made no effort to interview an individual accused by an associate of committing a serious crime.

Shifting focus from the individual to the institution, what should one think about a university that admitted students on the basis of the wealth of their parents; a newspaper that did not cover a scandal involving one of its chief advertisers; a scientific laboratory that suppressed data that discredited the value of a drug marketed by its principal funder; or an auditing firm that refused to take clients who espoused a particular religion?

While few observers might argue openly for it, many would resign themselves to the state of affairs I've just described. On their account, it is human nature to succumb to these lures. Human beings always have, and always will, pursue their own interests; and so will the institutions that, after all, are created, populated, and sustained by members of the same species. The most that we can hope for is that, in an open marketplace, such dishonorable individuals and flawed institutions might ultimately fail—perhaps because of negative publicity, poor management, or services less adequate than those provided by competitors who might be more, but possibly even less, ethical.

In what follows, I take a contrasting stance. I argue that at both the individual and the institutional level, it is possible and indeed desirable to adopt a *disinterested* stance. In assuming such a stance, the individual, group, or institution deliberately brackets personal preferences and the possibility of imminent rewards. Rather, the disinterested stance entails a commitment to judge each particular case on its merits; to consider the full range of options; and to select the course of action that embodies the longstanding values of the profession and thereby serves the common good over the long haul.

Accordingly, in the individual cases just mentioned, the physician recommends the treatment that is most appropriate for the patient; the teacher divides her time roughly evenly across students; the auditor applies the same standards to each and every company whose books he examines; the judge recuses herself at the slightest hint of a conflict of interest; and the journalist makes every effort to interview the accused as well as the accuser.

The same kind of disinterested stance can and in the ideal world should be adopted at the institutional level. The university has need-blind admissions; the newspaper fully covers all significant scandals; the scientific laboratory publishes all of its findings; the auditing firm applies the same standards to all clients.

In principle, a disinterested stance can be assumed in any sector of life—parents toward their several children, a movie director toward all members of her cast, a business toward the full range of suppliers or customers. As suggested by the examples cited in the opening paragraphs, my focus here falls on the professions—on the individual professional and on organizations staffed primarily by professionals. In speaking of professionals, I mean to single out those individuals who have received extensive training followed by official certification; have explicitly or implicitly pledged to honor longstanding values of the profession; are empowered and trusted to make complex decisions under conditions of uncertainty; and can be expelled from the ranks of the profession by individuals or agencies so empowered (Abbot, 1998; Freidson, 2001; Gardner & Shulman, 2005).

I contend that if its citizens are to be served fairly and comprehensively, any society must both cultivate and protect the professions. Historically, the launching and development of the professions is a long and arduous task; in contrast, the *dissolution of a profession* can be accomplished with relative ease and remarkable speed. The advent and—increasingly—the hegemony of the digital media complicate the process and practice of professionalization. And yet, unless we are willing to dispense totally with the disinterested stance, we need to devise ways to perpetuate the ensemble of professions, the niche of the individual professional, and the operation of institutions staffed by professionals—the vital preserves of disinterestedness.

At both the individual and the institutional level, one can discern times and circumstances when the disinterested stance emerges. With respect to the individual, in early childhood the young person—necessarily in the clutch of an egocentric perspective—thinks only about his needs and desires. By the time of formal schooling, however, most children have adopted a sense of fair play—goods ought to be evenly distributed, cheating undermines the rationale for a game.

With respect to the broader society, the emergence of legal systems over the millennia undermines sheer assertions of power, replacing them with a legal system ("a government of laws, not of men, or women") and with judges who can administer penalties in an even-handed manner. One can point to specific codes—such as the Hippocratic oath in medicine or the ethical guidelines of journalism—that attempt to identify overarching values, clear spheres of interest, and specific practices that counter or minimize the pursuit or the protection of parochial interests.

To be clear, the emergence and the crystallization of a disinterested stance is never simple and straightforward. Human beings are all too capable of believing that they can subordinate their own interests, though it is apparent—at least to disinterested observers!—that they (or we) are often deluding ourselves (Ariely, 2012). Newly instituted governments may decide that it is in their interest to eliminate—or at least reduce the power of—independent, noncompliant sectors. Indeed, professions rarely survive in recognizable form during totalitarian or authoritarian regimes, be they regimes on the far right or on the far left. And repeatedly over the course of history, new technologies disrupt the professional landscape.

Consider, for example, the effect of the invention of printing on the professoriate at universities. Whereas teachers could once have put forth their own views as if they represented a universal consensus, their students could now consult books that foregrounded views that had earlier been marginalized or suppressed. Turning to another domain, it is difficult to think of the practice of science in the absence of ready access to published accounts of systematic observations and well-controlled experiments, especially ones that challenge the conventional wisdom (or stupidity). The new digital media may constitute an even greater disruptive force on current conceptions and practices of disinterestedness.

## TERMINOLOGY

Even for those who understand that "disinterest" does not mean "lack of interest," the term can pose difficulties. A disinterested stance is not antithetical to strong motivation or feelings; indeed, these features may even be a reliable concomitant of the dedicated professional. The disinterested journalist may have a passion for justice; the disinterested educator may love teaching and research; the disinterested scientist is often motivated both by a strong interest in the topic that she is exploring and by the aspiration of making a fundamental discovery.

I've deliberately chosen the term "disinterested." Unlike the more common "fair play" or "just," it does not carry excessive connotative baggage. Nor is it simply *neutrality* or *objectivity*. The term "disinterest" presupposes that all of us have interests, but that it is an essential part of the role of the professional—whether officially defined or simply aspirational—to be willing and able to put those interests aside.

What marks the disinterested stance is the capacity to separate out, to assume a distance from, one's own personal motivations—or, to put it positively, to prioritize instead a passion for the practice of the profession at the highest level of excellence. And so the investigative journalist is prepared to abandon pursuit of a promising story if the supposed facts turn out to be illusory; the teacher does not feature his personal point of view on a controversial issue but gives a rounded picture of the topic and makes sure that opposing perspectives are fairly introduced; the scientist is willing to publish results even if these findings undermine a theoretical position of which she has previously been a staunch advocate; the lawyer defends with equal vigor the client with whose personal philosophy she is sympathetic and the client whose personal philosophy she finds repugnant.

Taking a leaf from sociological analysis, it is important and perhaps necessary to think of the professional as an individual who assumes a particular *role*. In a way analogous to the surgeon putting on a surgical gown or the judge donning a robe, the professional takes on a *set of attitudes and standards* when he or she is practicing that profession. It becomes vital for the professional to be cognizant of when she is assuming that role; to make it clear as well to observers; and, most challengingly, not to confuse that role with other roles. Those professionals who attempt to assume more than one role—for example, the journalist who is also a lawyer or a teacher—require keen self-awareness and, on occasion, self-censoring.

Time for a few examples. During the 1980s and 1990s, Maine Senator George Mitchell was clearly a strong advocate of the program of the Democratic Party. More generally, he regularly received very high ratings from liberal watchdog groups and correspondingly low ratings from their conservative counterparts. Clearly, Mitchell was not devoid of values or passions.

Yet year after year, even after he became Senate Majority Leader, Mitchell was cited as the most respected senator by aides from both parties. Moreover, at times when the United States or other countries were facing a difficult situation—the use of steroids in American baseball, the decades of bloodshed in northern Ireland—Mitchell was chosen as a mediator by the opposing parties or "interest groups." Before becoming an elected member of Congress, Mitchell had been a federal judge, and as a judge, he had learned to put aside his own pet interests and to deliver fair verdicts and sentences. This capacity to act in a disinterested manner, to assume the role of *judge*, was the reason that embattled parties, time and again, turned to Mitchell.

Daniel Schorr was for many years a leading journalist on radio and television. He seems to have been bitten by the journalism bug at an early age. Schorr had his first scoop at age 12, when a neighbor fell to his death. He recalls,

I called the local newspaper the *Bronx Home News*, and dictated a story to them, earning my first five dollars . . . And what I remember of it was that my mother and others remarked about how cool I was, how unaffected I was myself, emotionally, by the fact that a person had died almost in front of my eyes. That anecdote, whether literally true or not, has colored my entire professional life, the ability to detach myself.

That capacity for disinterestedness was severely tested when Schorr was assigned to do a documentary about the Nazi death camps. In his own words:

It was pretty strong stuff at Auschwitz . . . There was a time, for example, when I was saying in my script, and this is where they came out of the gas chamber and then they pushed them into the ovens over there . . . they couldn't get all the people they were killing, so some of them were just burned from the oven and thrown into empty trenches over here . . . I did that very journalistically. I didn't faint. I didn't feel overcome by it all, I just did my job. It may well be—there were members of my family two generations ago who were lost in the Holocaust. . . . Some of them may well have died at Auschwitz, but I was doing a job there and I did the job. (Shorr, D. quoted in Gardner et al., 2001, p. 181)

Note that neither Mitchell nor Schorr were individuals devoid of passions or feelings. They could not have done their jobs well if they took a passive or perfunctory stance. And yet when called upon to fill their *professional roles*—respectively, of judge and of journalist—they seem to have been able to put aside personal preferences and feelings and to proceed in a disinterested fashion.

Disinterestedness need not depend solely on the judgment and the courage of the individual practitioner. Individual lawyers may give their all to a client for whom they have sympathy, because they can count on the judicial system—the opposing lawyer, the jury, the judiciary, the appeals process—to provide, when needed, a counterweight. It is the job of the editor of the paper to make sure that an important story is covered by more than one reporter . . . and that biases in reporting are balanced or, better, edited out by the time the media outlet goes to press. A dean or provost should make sure that across the faculty of a department, rival points of view are expressed. Perhaps most powerfully, the entire apparatus of scientific funding, publishing, and awards is set up to make sure that faulty claims are recognized as such—and as soon and as publicly as possible.

Dependence on "the system," or on higher authorities, can only go so far, however. A liberally oriented newspaper may feel that it can push its own agenda because there is a rival paper on the conservative side. But if one of the newspapers ceases publication, it is no longer possible for readers in the region to receive a balanced picture of events. A judge in a lower court may rely on higher courts to correct his excess, but there is not an infinite regress of courts on which to depend. Ultimately, the buck must stop somewhere; it is preferable that each level of the system—in the United States, from district courts to the Supreme Court—exhibit disinterestedness.

I like to invoke the metaphor of a righting mechanism. Without doubt, both individuals and institutions have enormous potential to pursue their self-interest and nothing more. It is important for both individuals and institutions to be aware of this tendency and to guard against its exploitation. As a liberal Democrat, Senator Mitchell certainly had pet policies as well as pet peeves. But when he was asked to mediate among three competing bills, the proponents looked to Mitchell not to push his own interests but rather to come up with a formulation that, while not perfect, would be judged as satisfactory and fair by the several parties. As a Jewish person who had lost relatives in the Holocaust, Daniel Schorr certainly had strong feelings about the Nazi Holocaust; and yet when called to report on the death camps, he controlled his emotions and presented the facts in a straightforward way.

As I see it, members of every profession should remain vigilant that disinterestedness remains a treasured value. In the current cable television news arena in the United States, MSNBC and Fox News deliberately assume rival partisan positions, while PBS styles itself as a moderate middle (if not explicitly disinterested) voice. Clearly, should one of these networks cease operation, another "righting mechanism" (or, perhaps, "lefting mechanism") would be needed.

Interestingly, the disinterested stance is not always appropriate. For example, there is no reason why a painter or novelist should hide his or her passions; nor do we look to such artists for a disinterested presentation. At most, we expect the artist to present his own perspective, not that of others. ("Precedence" is a hallowed precept in the law, something often to be spurned in the arts.) An athlete should do his best to win a competition—though we have the right to expect the athlete not to cheat and, to the extent that he is involved in team sports, to bracket his personal glory in service to the team. Individual businessmen as well as corporations properly work for profit. We admire those that treat their customers well and that work for the welfare of the broader community—but strictly speaking, such disinterestedness is not part of their role, not part of their job description. No entrepreneur can be barred from practice simply for pursuing maximum profits.

One can certainly take the position that it is unnecessary or unrealistic for ordinary citizens to attempt to be disinterested; rather, they should become informed and then advocate strongly, if also fairly, for the position that they personally favor. I can accept this stance as a fallback position—after all, it is better to be informed than uninformed, and it is better to be an advocate than to be silent or hypocritical.

Finally, it is possible for disinterestedness to go too far. Therein lies the peril of bureaucratization. Bureaucracies were established for the legitimate purpose of moderating the pursuit of special interests. Cases are judged not by the power or persuasiveness on the part of individual practitioners, but rather according to a "neutrally" established set of rules. Who gets a driver's license, and in what order petitioners are seen, should be done in a transparent and evenhanded way.

But what sociologist Max Weber caustically termed the "iron cage" (1958) of bureaucracy can be counterproductive. In cases where complexity or subtlety are in order, the rule book may be inadequate or lead to results that are widely seen as destructive or as benefiting the wrong party. Here a genuine recognition of competing interests, and how they can most judiciously be weighed against one another, trumps lockstep adherence to an algorithm. Disinterestedness can go too far! King Solomon is sometimes to be preferred over the Official Manual of Rules and Regulations.

## THE DIFFERENCE THAT A HALF-CENTURY CAN MAKE

The disruptive potential of the digital media can clearly be seen by a comparison with the state of affairs that obtained in the United States (and perhaps elsewhere) more than a half-century ago. At the time, as expressed by an editorial in the journal *Daedalus*, the professions were at their height of power and influence. Editor Kenneth Lynn exclaimed, "Everywhere in American life, the professions are triumphant" (Lynn, 1963). Many if not most professionals saw themselves as acting in the public interest; rather than attempting primarily to amass individual wealth and status, they believed that they had entered into an arrangement with society. If they were afforded prestige and a reasonably comfortable living, they would in turn attempt to render complex, fair judgments under conditions of uncertainty.

It must also be noted that in the middle of the last century, the professions were overwhelmingly dominated by white males, typically WASPS (White Anglo-Saxon Protestants). That the professions were virtually closed to many sectors of society is a strong mark against them but should not be allowed to obscure their positive, more public-minded facets. More generally, in reflecting on this earlier era, we should avoid nostalgia for a golden age that certainly did not exist for all parties; nor should we throw out the "professional baby" with the often "prejudicial bathwater."

Even today, while laudably open to a much broader swathe of society, it could be maintained that the professions remain bastions of privilege. Credentialing certainly remains; credentials from elite institutions are as coveted as ever. Professionals also do well financially. Yet hardly any profession is held in anywhere near as high regard as it once was. Talented young persons who might at an earlier time have become professionals are now attracted by careers as investment bankers, management consultants, creators of new media or technologies, or serial entrepreneurs—careers where they not only avoid the glare of publicity but where—even as youths—they can amass huge fortunes. Note that while many of these new occupations adopt the lexicon of *professions*, they do not have the credentialed status, nor (in my view) do the members of these new professions routinely behave as genuine professionals.

For their part, even fully credentialed professionals today are rarely seen as individuals acting in the public interest. Rather, they are widely regarded as rank-and-file employees of huge institutions. Often these increasingly corporate entities are managed by individuals drawn from the realm of business, rather than from the very professions over whose members they hold sway. (See Essay 30 on the sudden demise of the law firm Hill & Barlow.)

## ENTER THE DIGITAL MEDIA

There is little question that the professions—and their vaunted if not always achieved disinterested stance—have become exceedingly vulnerable in the time of the digital media.

Several factors are responsible. To begin with, members of any developed society now have direct, immediate access to a huge amount of information on nearly every conceivable topic. In the past, professions and professionals benefited from an aura of mystery—merited or not—surrounding their expertise. While perhaps not as secretive as medieval guilds, few individuals outside the professions believed that they had the knowledge and wherewithal to challenge the opinions of the certified, credentialed experts. The credentialed professionals looked authoritative, and they spoke (and dressed!) with authority.

In addition to greater access to the knowledge and skills once restricted to expertise, there is far greater knowledge about the weaknesses and vulnerabilities of individuals and institutions. In the era of Franklin Roosevelt, few Americans realized how physically impaired their President was; and while known to supposedly disinterested reporters, the sexual escapades of John F. Kennedy (and other highly visible political figures across the political spectrum) escaped public attention and scrutiny.

Now virtually every peccadillo of virtually every public figure—whether true and verified, true but not adequately verified, suspected, rumored, or simply spun out of whole cloth—is available for public scrutiny. Attention to improprieties by individuals—be they senators, scientists, surgeons, or scholars—inevitably colors—if, to switch metaphors, it does not poison—the way the sector to which they belong is viewed by the general public. Even if the majority of auditors treat all clients in a judicious and even-handed way, the whole profession is tainted by the well-known malpractices of Arthur Andersen and other gigantic accounting firms.

Perhaps most damaging to disinterestedness in the professions, in the long run, are alternative forms of credentialing. In the 20th century it was widely believed that aspiring professionals benefited from personal apprenticeships, in which they spent considerable periods of time in the field working under or alongside acknowledged and respected mentors. These apprenticeships ranged from internships and residencies for junior physicians to stints in regional bureaus for cub reporters to extensive benchtop work by budding scientists in laboratories presided over by highly credentialed and well-established researchers (Nakamura et al., 2009).

With the ubiquity and accessibility of the digital media, the special status of credentialing has become extremely vulnerable. This situation is well captured by the famous cartoon with its witty caption, "On the internet, no one knows that I am a dog." Not only is it possible for new entries and entities to claim that they can provide equivalent training online—from a bachelor's degree to a professional degree—but many consumers will be unable to separate the professional wheat from the bogus chaff.

At least in principle, one could implement the professional equivalent of the Turing test—a well-known thought experiment in computer science. If a hypothetical graduate of an online law school can pass as—or for—a graduate of Yale Law School, or an online physics doctoral candidate can pass for a PhD from MIT, what possible justification can there be for multiyear training far from home at great expense? Perhaps we can now say, "On the web, no one knows—and perhaps no one cares—that I've never actually set foot in a bona fide, credentialed medical school."

In a day when virtually everything is counted and ranked by supposedly neutral search engines that can all too readily be gamed, it becomes extremely difficult for all but the most informed to make needed distinctions among claims of quality. In principle, to be sure, one should be able to distinguish among (1) genuine expertise and knowledge; (2) the ability to parrot such knowledge, thanks to good briefing or "test practice"; (3) skillful branding or advertising; (4) knowledge of how, or the good luck, to "go viral"; and (5) credentials that are in fact valid as opposed to those that are suspect or even invented out of whole cloth.

But how many of us have the time, the motivation, and the expertise to undertake such needed discriminations? And how many ranking systems obscure rather than illuminate the most important criteria of quality?

While it might be tempting to see the digital media as a villain, such a judgment is inappropriate. Indeed, technologies in themselves do not have interests—they can even be described as the ultimate disinterested entities. A pencil can be used to write exquisite sonnets or to poke out a competitor's eyes! By the same token, a website can be used to raise money for victims of a natural disaster or to fuel a campaign of racial hatred.

But while technologies do not have interests, they certainly have biases and tendencies.

In comparison to earlier media, the digital media are essentially instantaneous, interactive, and open to immediate editing or refuting, enabling virtually unlimited networking and connectedness. Search engines have to be programmed, their parameters can be changed, and indeed different search engines will come up with different entries and priorities. In the incarnation with which we are currently most familiar, the digital media also favor messages that are short, sharp, flashy, and easily transmitted and transformed with the potential to go viral, with results that may or may not be predictable or desired.

To be blunt, none of these tendencies initially favors a disinterested stance. After all, the disinterested stance tends to be fostered and found among those with lengthy professional training, rather than among those who may have little knowledge or depth, but much to say or even to shout. Or, more poignantly, "on the Internet no one cares that I have a PhD, two postdoctoral fellowships, and favorable recommendations from respected professionals." Indeed, nowadays, if an aspiring physician wants to gain attention or even clients, she may be better advised to do an internship on Madison Avenue or Fox News, or to advertise in the pages of a flight magazine, rather than to pursue a residency at Memorial Sloan Kettering or Massachusetts General Hospital.

Still, whatever their apparent initial proclivities, there is no reason in principle why the digital media *need* to deal a death blow to disinterestedness. For one thing, we live at a time when—at least in principle—it is *more possible* than ever to determine what actually happened or did not happen. Individuals with the time and energy to scan the web are in a far stronger position to discover the actual state of affairs. Whether one approves of Wikileaks or Wikipedia (and they are not the same!), these sites dramatically increase the probability that all possible information (and misinformation) has been posted.

An additional point is that disinterestedness need not inhere in the media themselves. The media might be better construed as playgrounds on which opposing perspectives can and should be articulated, as powerfully as possible. It's not the job of the web to mediate; it is the job of the web to make all manner of information available. Then—and, some would add, *only* then—individuals or institutions can adopt a disinterested stance, if that is their preference.

Of course, the tendency of individuals to examine chiefly the sources with which they are likely to find themselves in agreement has been well documented (Bishop, 2008; Sunstein, 2011). But even though some algorithms direct us to sites with which we are likely to agree, we cannot simply blame the media for such selective presentations; ultimately, bias is chosen (consciously or unconsciously) by individuals—and it can be countered by effective education and a healthy dosage of self-discipline. As an expert on the topic, I certainly know which sites on the topic of human intelligence are likely to present one point of view, which are likely to embrace the opposite perspective, and which seem to make a determined effort to provide broad coverage and strive for a disinterested stance.

Paradoxically, the very plethora of information available on the digital media could usher in a new era of disinterestedness. Given the deluge of information, which no human being can fully absorb, we may end up cherishing *especially* those individuals, groups, and sites that manage, nonetheless, to transcend individual and group interests.

If I want to know what is going on in the world in 2024, I rely on a few broadcast outlets—National Public Radio and the BBC—and a few publications—*The*

*New York Times*, *The Economist*—to provide a wide ambit. Perhaps each of these outlets has slight biases as well as a distinctive cultural orientation, but these leanings are well-known. Importantly, when these media outlets are alerted to errors, the editors make every effort to correct the error promptly and prominently.

At times, the new media can aid in disinterestedness. Consider the example of athletics. Until the invention of speedy recording and transmitting devices, close calls were left to the discretion of the judge or referee. Even when the judge strove to be disinterested, it was quite possible for him to err. Now, however, thanks to rapidly accessible digital records, it is possible to examine the questionable play and subsequent call quickly and to make amends, as appropriate. No longer is it a case of "It's only how I call them"—but rather a case of "It's how I call them after I've had a chance to review the instant replay."

Less visible, but perhaps more important, are the individuals who administer complex informational systems. Because of technical problems that arise, such individuals may have the power to read users' emails, to access credit information, and to follow the trail of sites visited and deals made. When sticky ethical issues arise—for example, involving individual privacy in the face of requests by governmental agencies—these administrators often work in tandem with the chief officers of highly trafficked sites like X (formerly Twitter), Facebook, or Google.

To this point, it appears that individuals charged with these huge responsibilities have conducted themselves with admirable professionalism. According to Ethan Zuckerman, "when systems administrators go rogue and read email, they tend to get shunned and drummed out of the profession" (personal communication, July 31, 2012). It may be that the importance and delicacy of their situation serves to engender an ethos of professionalism in the ranks of such administrators.

I should note the claim that there is "wisdom in crowds." As a general principle, I find this cliché unconvincing. There is a huge difference between conclusions reached by individuals who lack knowledge or who act in their own self-interest, as compared to conclusions reached by individuals who are deeply informed and who engage in deliberative reflection. I shudder to think about what would happen if policy decisions on climate change, or foreign aid, or separation of church and state, or the teaching of science, were made on the basis of plebiscites about intelligent design, the existence of angels, or the allocation of the federal budget.

Reliance on expertise may not always yield superior results (experts are also susceptible to groupthink)—but knowledge and serious deliberations are necessary, even if they are not sufficient. Moreover, while experts are certainly not immune from bias, they are at least likely to be countered by others of equal expertise. In contrast, history is replete with examples of manipulated masses, culminating in the madness of crowds.

I've argued that even before the hegemony of digital media, the disinterestedness once associated with individual professions and with professional guilds had been attenuated. With little question, the complex of Internet, web, social network, quick and brief messaging, and other digital entities has contributed to the undermining of the professions and the dissolution of the disinterested stance that had once been key to their achievements. If we are not to give up on the ideal of disinterestedness, what can we do and what should we do?

A number of levers come to mind. First of all, we need to bring to the attention of various publics the importance of disinterestedness and the perils that will fall—indeed that have befallen—those sectors where disinterestedness has waned or even disappeared.

Though not myself a marketer or brander, even I know that the word "disinterestedness" is a nonstarter. Phrases like "fairness," "even playing field," "unbiased information," "a good umpire," or "an impartial judge" make the same point in a more popular vein. Quite effective are movies and television shows featuring people who do the right thing—blowing the whistle on unethical behaviors and instead modeling what it's like to be fair-minded. Perhaps it's time for a movie about a highly ethical systems administrator!

As a complement, we need to highlight examples of *lack* of disinterestedness—to turn an even less mellifluous phrase, the costs of "dis-disinterestedness." Only when individuals who pervert the values of a profession are publicly chastised—only when there are well-publicized wake-up calls—is there a reasonable chance that others in the profession will mend their ways.

I certainly don't favor the reinstitution of public stockades. But unless there is some kind of public recognition, some kind of shaming of those who have pledged to be disinterested but choose not to be, it is hard to see how young workers are ever to appreciate the sheer cost to the profession of conduct that flaunts disinterestedness. Put sharply, it matters whether bankers who play with interest rates (or traders who manipulate the cost of energy) are protected by U.S. regulatory or legal entities; allowed to resign with full pay; forced to resign and to return their bonuses; indicted but get away with a slap on the wrist; or indicted and, if convicted, given long, well-publicized jail sentences.

Urgently, we need to create spaces—both face-to-face and online—where members of a profession can discuss the difficult cases that pop up with regularity; what steps they have taken in response to these quandaries; what worked well; what might have been planned and executed in a more effective way; and how to draw on the intelligence distributed across the community so as to increase the likelihood of disinterested decisions and actions.

Drawing on a figure of speech with a lengthy history in the Anglo-American world, I call such spaces "commons." Common spaces developed originally as pastureland in which the cattle of neighbors could graze. In the contemporary era, we should create common areas—virtual as well as physical—where members of a guild can meet to discuss the most important and most difficult cases that arise. (One example at the time of this writing is the blog The Volokh Conspiracy, where law professors and others discuss complex cases and vexed issues. Another quite different example is the experiment in the evaluation of patent applications described by Beth Simone Noveck [2009]. For an excellent discussion of the vital role that common spaces can play, see the writings of Nobel laureate Elinor Ostrom [2009].)

My colleagues and I have been engaged in such work. When we first began the Good Work Project in the mid-1990s, we were concerned primarily with the question of how professions can survive in light of a number of disruptive conditions: the relentless pace of change; the altering of our sense of time and space by the digital media; and the power of unregulated markets. We did not use the term "disinterestedness," but we already had a sense that the professions harbored important human qualities and values and that these had become exceedingly vulnerable in the current era. We came to value the nature and importance of work that is technically excellent, personally engaging, and carried out in an ethical manner (see Essays 28–31). For "ethical," I would happily substitute the descriptor "disinterested."

While most of our work has been with young persons, let me say a word about how we have approached professionals on our project.

One group with whom I worked were school principals. I began by asking, "To whom or what do you feel responsible?" We then reviewed the various interest groups in the school and how best to navigate across their often-competing agendas. I described dilemmas faced by a principal: for example, what to do when a teacher of high quality is a very tough grader; her students, while appreciating the teacher's pedagogical gifts, feel that she is jeopardizing their entry to college; or what to do when the offspring of the chair of the school board is caught cheating.

Such dilemmas arise almost every day. Some of them may be sufficiently familiar that the principal has hit upon a comfortable and effective course of action. But others are new and complex, and it is difficult to figure out the optimal response. Indeed, many of these dilemmas involve the digital media: Should students be allowed to text during class, should students be allowed to "friend" teachers on a social network, what happens when answers to an assignment go viral? Here some kind of common space, in which rival positions are articulated and scrutinized, could be quite helpful.

Working with law students, I began by asking, "In whose interest(s) should a lawyer work?" I then asked whether the supreme judicial court of the state (or of the nation) also had interests and, if so, what they were, and what it would mean if no members of said court were able to behave in a disinterested fashion. Drawing on a newspaper account, I then described a meeting in New Orleans of lawyers who work for large corporations. At this meeting, the lawyers admitted that they were frequently faced with conflicts of interest—for example, being offered a huge bonus by a corporation if a case turned out in the desired way. I quoted further from an article: "Lawyers—even Wall Street lawyers—are supposed to be a different breed from their clients and banker counterparts. They sign an oath when they are admitted to the bar. Their ethical standards are supposed to be beyond reproach. For many years, if a lawyer was called 'commercial' that was considered pejorative. Today it is increasingly a badge of honor. Indeed, much of the conversation among the top lawyers in New Orleans was not about the rules but about how far you could push them" (Sorkin, 2012). After my talk, the future lawyers lingered in the hallway for quite some time, discussing the pros and cons of a life in the contemporary practice of corporate law.

It is worth considering the content of a lawyer's oath. Here are some excerpts from the lawyer's oath in the state of Michigan:

> I will not counsel or maintain any suit or proceeding which shall appear to me to be unjust, nor any defense except such as I believe to be honestly debatable under the law of the land.
>
> I will employ for the purposes of maintaining the causes confided to me such means only as are consistent with truth and honor and will never seek to mislead the judge or jury by any artifice of false statement of fact of law.
>
> I will accept no compensation in connection with my client's business except with my client's knowledge and approval.
>
> I will in all other respects conduct myself personally and professionally in conformity with the high standards of conduct imposed upon members of the bar as conditions for the privilege to practice law in the State. (*Lawyer's Oath*, State Bar of Michigan)

It would be interesting to verify how many lawyers in Michigan have been disbarred and what reasons, other than sheer illegal behavior, have ever been grounds for disbarment. (Google offers no ready answer to this question.)

Life is complex, situations are tricky, and no professional should hold him or herself up as a paragon of virtue, let alone an icon of disinterestedness. Indeed, such judgments are best left to others and are best made after the passage of time. Nonetheless, I feel confident in asserting that *any* profession and professional group can benefit from a discussion of the various interests at play in their work; that such discussions are likely to reveal how best to serve wider interests of the profession and the broader values of the ambient community; and that relevant discussions can and ought to be initiated well before the beginning of professional education.

While in certain ways this challenge is compounded by the hegemony of the digital media, the media can also be mobilized to advance the causes of good professional behavior. While a physical commons is necessarily limited to those who live in the neighborhood, a virtual commons can be opened up to interested practitioners and professionals around the world. If well and fairly moderated, such a commons may well come up with examples and rationales that help to upgrade professionalism and move toward a stance of disinterestedness. It is up to those of us with professional opportunities and obligations to ensure that the media are mobilized for the ends that we most cherish.

# REFERENCES

Abbot, A. (1988). *The system of professions.* University of Chicago Press.

Ariely, D. (2012). *The honest truth about dishonesty.* HarperCollins.

Bishop, B. (2008). *The big sort.* Mariner Books.

Freidson, E. (2001). *Professionalism, the third logic.* University of Chicago Press.

Gardner, H. (2012). *Truth, beauty, and goodness: Educating for the virtues in the age of truthiness and Twitter.* Basic Books.

Gardner, H., Csikszentmihalyi, M., & Damon, W. (2001). *Good work: When excellence and ethics meet.* Basic Books.

Gardner, H., & Shulman, L. (2005). The professions in America today: Critical but fragile. *Daedalus, 134*(3), 13–18. http://www.jstor.org/stable/20027993

Lynn, K. (1963, Fall). Introduction to issue on The Profession. *Daedalus, 649.*

Marshall, P. (2004). *Facing the storm: The closing of a great law firm.* http://www.goodworkproject.org/wp-content/uploads/2010/10/34-Facing-the-Storm-9_04.pdf

Noveck, B. S. (2009). *Wiki government: How technology can make government better, democracy stronger, and citizens more powerful.* Brookings Institute Press.

Ostrom, E. (2009). "Beyond Markets and States: Polycentric Governance of Complex Economic Systems." Nobel Prize Lecture, December 8.

Sorkin, A. R. (2012, March 13). Conflicted, and often getting a pass. *New York Times.* https://archive.nytimes.com/dealbook.nytimes.com/2012/03/12/conflicted-and-often-getting-a-pass/

State Bar of Michigan. *Lawyer's oath.* http://www.michbar.org/generalinfo/lawyersoath.cfm

Sunstein, C. (2011). *Going to extremes: How like minds unite and divide.* Oxford University Press.

Weber, M. (1958). Bureaucracy. In H. H. Gerth & C. W. Mills (Eds.), *From Max Weber: Essays in sociology* (pp. 196–244). Oxford University Press.

Zuckerman, E. (2012, July 31). Personal communication.

# The Future of the Professions

*For as long as I can remember, I have looked up to individuals who are called "professionals"—doctors, lawyers, and architects, to name a few. I aspired to be a professional, I became a professional, and I continue to hope that young persons of promise whom I encounter will become professionals.*

*But in recent years, I've become increasingly uncertain about whether professions will continue to exist, at least in a form that I can admire or even recognize. There are many views about why professions are on the wane—indeed, to put it sharply, whether the professions are being murdered or whether, to maintain the metaphor, they have been committing suicide. I'll deliver my interim judicial verdict at the end of this essay.*

*Since my own life so closely parallels the ups and downs of the professions in the last several decades, I begin in an autobiographical vein. Then, building on the arguments about the importance of the professions and their fragility (see Essays 28–32), I turn to an interchange with two thoughtful scholars who are quite skeptical about the aura that has traditionally surrounded this sector of society.*

## AUTOBIOGRAPHICAL REFLECTION

My parents, Hilde and Rudolph Gaertner, escaped from Nazi Germany in the nick of time, arriving in New York Harbor on November 9–10, 1938—the infamous "Night of Broken Glass." Many of their relatives and friends were not so fortunate. My parents were not themselves professionals—my father had been a businessman, and my mother's desire to be a kindergarten teacher had been thwarted by the rise of the Nazis. With neither professions nor funds, they soon found themselves living in very modest circumstances in Scranton, Pennsylvania, where I was born 5 years later.

Parental aspirations for my younger sibling, Marion, and for me were high. As I was the proverbial "bright Jewish boy who hated the sight of blood," almost everyone who thought about it (including me) assumed that I would become a lawyer. I went to Harvard College, where indeed I took a course in the law with constitutional scholar Paul Freund, who encouraged me to go to law school; but not one to leave any stone unturned, I also took some pre-med courses and, in the summer before my senior year, arranged an interview with the Dean of Admissions at Stanford Medical School.

Only after graduating from college and spending a year abroad in London on a fellowship did I make what was for me a daring decision: to abandon the prototypically aspirational professions and instead to pursue graduate studies in developmental psychology. For 15 years after receiving my doctorate, following a path that in retrospect was quite risky, I pursued full-time research on "soft" (research grant) money. In 1986, I was fortunate to receive a professorship at the Harvard Graduate School of Education. At last, I was a professional—indeed, a professor!—and our family's ambitions were fulfilled. And so I've remained until this time.

I've allowed myself this autobiographical indulgence because it closely tracks the course of the professions over the last several decades. When I was growing up, at the very time when postsecondary education was expanding greatly in the United States, becoming a professional (which did include becoming a college professor) was a very high aspiration. It was not, however, equally open to everyone. For historical and cultural reasons, most professionals were white, Anglo-Saxon males, principally from privileged backgrounds. If you were an immigrant or the child of immigrants, Jewish, female, of impoverished circumstances, and/or non-white, your career choices were restricted. Supreme Court Justices Thurgood Marshall, Clarence Thomas, Sandra Day O'Connor, and Ruth Bader Ginsburg merit admiration because of the overwhelming initial odds that they overcame! Fortunately, by the time that I was choosing a career, these barriers were receding, more quickly for the first listed groups than for the latter cohorts.

While I was in college, *Daedalus*, the publication of the American Academy of Arts and Sciences, published an issue devoted to the professions. As the editor Kenneth Lynn (1963) phrased it, "Everywhere in American life, the professions are triumphant." And indeed, my close friends and I almost all chose to enter one or the other of the professions. When, at our 50th college reunion, we jointly reflected on our choices of career, 10 of 12 were either doctors, lawyers, or professors—and the 11th had actually completed law school but had not practiced law (the 12th had become a movie director).

If we had polled the same individuals, and other of our classmates, about the careers that our children (and grandchildren) were pursuing, a far lower percentage would have responded with the traditional professions. Instead we would have heard repeated references to Hollywood, Silicon Valley, Wall Street, and phrases like start-ups, venture capital, angel investors, hedge funds, Google, Facebook, and Amazon (the latter started, we should note, by young persons who had attended, respectively, Stanford, Harvard, and Princeton). Few writing today, whether in *Daedalus* or the *Huffington Post*, would describe the traditional professions as aspirational, with respect to the options considered by the most sought-after graduates of elite institutions. And if young graduates did elect to pursue the traditional professions, they would quite possibly work at one or more of the aforementioned corporations.

## INTRODUCING THE SUSSKINDS

I now turned to a recently published book, *The Future of the Professions* (2015), by the British father-and-son team of Richard and Daniel Susskind. Richard has long been a leading thinker about and critic of the legal profession, and his son Daniel teaches economics at Oxford.

These authorities firmly believe that what has happened—and what is going to happen—to the practice of law will happen as well in the other major professions. They outline major technological innovations and training regimens for paraprofessionals, which in combination are likely to satisfy the bulk of the needs of ordinary clients as well as or better than do the ordinary run of solicitors and barristers. In their considered view, contemporary professions are failing economically, technologically, psychologically, morally, and qualitatively, and by virtue of their inscrutability (p. 33). Or, as they put it colloquially, "We cannot afford them, they are often antiquated, the expertise of the best is enjoyed only by a few, and their workings are not transparent. For these and other reasons, we believe today's professions should and will be replaced by feasible alternatives" (p. 3).

What may surprise most readers—at least, it surprised me—is that not only the younger but also the elder generation welcome these trends. To be sure, the Susskinds do not go quite so far as George Bernard Shaw, who famously quipped that the professions are "a conspiracy against the laity." But the father-and-son team celebrate the many freeing and empowering facets of individuals taking greater control of their own lives; and they do not hesitate to critique the professions for their less attractive features—their arrogance, their secrecy, their elitism. The Susskinds counsel young people not to enter the traditional professions, a dying sector; instead they celebrate cohorts of knowledge engineers, process analysts, designers, system providers, data scientists, and systems engineers (p. 264). If, as they anticipate, the professions are headed for a single burial plot, or for a whole sector of the cemetery, the Susskinds will shed few tears.

As a professional and an observer of professionals, I certainly discern and sometimes welcome the trends described by the Susskinds. In my scholarly work, I benefit from the tremendous power of search engines and the ability to share and edit documents seamlessly. In my teaching, I communicate constantly with students and exchange all kinds of digital materials with them; and in class, I often use digital materials. Even before the advent of social media, I made all of my lectures available on the university's intranet. And of course, I can follow both my areas of special interest and what is happening in the world with great ease—and even comment on this material; in fact, this essay was posted initially on my website. I would not want to return to a predigital Dark Age.

As thoughtful analysts and skilled debaters, the Susskinds not only submit a strong brief in favor of their predictions, but also anticipate possible responses to their account. In a section of their book called "Objections," they list the principal reasons why others might take issue with their analyses, predictions, and celebratory mood. This list of counterarguments to their critique includes the trustworthiness of professionals; the moral limits of unregulated markets; the value of craft; the importance of empathy and personal interactions; and the pleasure and pride derived from carrying out what they term "good work." With respect to each objection, the Susskinds give a crisp response.

I was disappointed with this list of objections, each followed by refutation. For example, countering the claim that one needs extensive training to become an expert, the Susskinds call for the reinstatement of apprentices, who can learn on the job. But from multiple studies in cognitive science, we know that it takes approximately a decade to become an expert in any domain—and that presumably such a decade includes plenty of field expertise. Apprentices cannot magically replace well-trained experts.

In another section, countering the claim that we need to work with human beings whom we can trust, they cite the example of the teaching done online via Khan Academy. But Khan Academy is the brainchild of a very gifted educator who in fact has earned the trust of many students and indeed of many teachers; it remains to be seen whether online learning à la Khan suffices to help individuals—either professionals or their clients—make complex technical and ethical decisions under conditions of uncertainty.

The Susskinds recognize that the makers and purveyors of apps may have selfish or even illegal goals in mind. But as they state, "We recognize that there are many online resources that promote and enable a wide range of offenses. We do not underestimate their impact of threat, but they stand beyond the reach of this book" (p. 233).

Whether or not one goes along with specific objections and refutations, another feature of the Susskinds' presentation should give one pause. The future that they limn seems almost entirely an exercise in rational deduction and accordingly devoid of historical and cultural considerations.

Turning to history, in the past century alone, the world has witnessed two world wars, a Cold War, and a Holocaust as well as other genocides, and we live at a time of strife and conflict in many corners of the world. New problems are sure to arise: the possibilities of nuclear explosions; cataclysmic climate change; unplanned mass immigrations; widespread religious strife; occupational dislocations; unintended consequences of digital technology; and an ever-increasing elderly and disabled population. There is every reason to believe that existing and possibly new professions and professionals may well be needed to deal with these situations. The historical lens has been left in the drawer.

(Writing 10 years later, I would add that neither the Susskinds nor I anticipated the advent of large language instruments like ChatGPT. Of course, these may have disruptive powers that we could not have imagined—they may also catalyze, perhaps even necessitate, the emergence of new professions.)

As for cultural considerations, the professions have arisen largely in the Western hemisphere, and when they have been adopted in parts of the Far East and the Southern hemisphere, they inevitably take on their own coloration and nuance. We cannot determine whether, and, if so, to what extent, modern economic development was enabled by the status and disinterestedness of the ensemble of professions. We do know that in many parts of the world the Western approach to professions has been admired and emulated. In fact, research that I carried out with colleagues documents that immigrants to the United States may be critical of the professionals whom they personally encounter, but they appreciate that *at least in principle* American citizens can get a fair trial and an honest report of the news.

At the same time, there is hardly universal agreement about how professions should be implemented. Taking journalism as a particularly vivid example, there is little agreement across nations with respect to issues of privacy, anonymity, and criticism of the government. At a time when nationalism is everywhere on the rise, where religions continue to clash, it is naïve to think that a professional algorithm—conceived and fashioned in Silicon Valley or in Britain's Cambridge Cluster—will carry out its work with equal efficiency and equal effectiveness around the globe. And if each region has its own apps, who or what will mediate among them? Some kind of AI "super-app"?

Even in the developed world, with its ever-increasing inequities, there is much work to be done with respect to human goals (e.g., social welfare, medical care) and humane ways of achieving them (e.g., nurturing teachers, responsive civil servants, empathic health workers). Moreover, we should note the rise of "countercultural" phenomena. At the very time that digital devices are flexing their muscles, there is the ascent of the maker movement, where hands-on arts and crafts are cherished. Who can possibly predict the ultimate balance between human making, mechanical making, and digital making?

None of these considerations means, of course, that it can or should be business as usual for the professions. Technologies will alter the professional landscape, and deservedly so. But it is naïve to think that an account arrived at largely through reasoned argument suffices for anticipating the many possible futures of the professional landscape.

## A QUESTION OF VALUES

With respect to the professions, the stakes are high. At their best, the professions arose to meet the most basic and deeply desired human needs and expectations. In much of the world and in much of recent history, professionals have been the chosen means—and the chosen role models—for meeting those needs: aspirations for physical health, mental health, justice, safe buildings, equitable financial institutions, mastery of major scholarly disciplines, and nurturance of the "better angels" of the inhabitants of the planet. Again, at their best, those who assume the role of professionals represent the competent and humane ways in which individuals carry out these tasks; they serve as models for at least one kind of person to whom we can aspire and one kind of society in which we would want to live. May we never have a world wherein the phrases "she is acting professionally" or "he is a true professional" become anachronistic or devoid of meaning.

One could say that, in theory, these needs and roles could be satisfied in a completely marketized society—but we have plainly seen that they are not. Indeed, as amply documented by recent scandal, the avowedly marketized society does not even prevent economic disaster; we need to regulate markets firmly and fairly. One could say that, in theory, these needs and roles could be satisfied in a completely digital society. But bits can never yield singular values; like any tool, they can be used to promote harmony or discord, greed or selflessness.

Indeed, there is no such thing as a purely disinterested algorithm—as is dramatically demonstrated in examinations of the algorithms used and regularly modified by major tech companies. Humans create and rearrange the bits—but then the values of those creative human beings become crucial. And as humans, we are simply not smart enough to be able to anticipate just how those bits will be arranged. Social media can engender strong friendships, but they can also engender bullying, depression, perhaps even suicides. These media can build community, but as in the case of an anonymous message delivery system, they can rapidly destroy community. Valid values require constant human construction, surveillance, critique, reflection, rebuilding, and a commons in which we all have a stake, over which we all care, or, to put it more sharply, we all should care.

Consider a folktale that exists in several versions: There is an artistic object that is so beautiful and captivating that people come from all over to examine it. One day one of the admiring visitors enthuses, "What a magnificent chalice; it's the most amazing thing that I've ever seen." A cynic grabs a hammer and says, "You think that's amazing—I'll show you something even more amazing." With that, he strikes the object and it breaks into innumerable pieces, which then scatter to the wind.

It took many years and many people to construct the professional landscape in which many of us grew up and took for granted. The disruptive forces in our society—intentionally or simply by virtue of their existence—have the potential to slay, to murder the professions. Alas, both the greed for money and the belief in the infallibility of the market have contributed to this lamentable possibility—I consider these factors to be suicidal impulses found in too many professionals. It's high time for those of us who continue to value the professions to reinvigorate and, as necessary, reinvent the professions. We need to acknowledge our complicity in the current undesirable situation; embody the principles and values that have enabled professional practice at its best; and work to ensure that they will be strengthened, not undermined, by the technologies to come, and insofar as possible, in harmony with the ever-unpredictable winds of history and culture.

## REFERENCES

Lynn, K. (1963, Fall). Introduction to issue on The Profession. *Daedalus*, 649.

Susskind, R., & Susskind, D. (2015). *The future of the professions: How technology will transform the work of human experts*. Oxford University Press.

# MINDS OF THE FUTURE

# Five Minds for the Future

## An Introduction

*By training and inclination, I am basically a researcher: I like to contemplate puzzles, figure out how to gather relevant data, analyze the data, reach conclusions, and then move on.*

*But as one ages, and particularly in the arena of education, one is often asked, "What should be done?" And so, early in the 21st century, I began to focus on this important and challenging question.*

*The answer that I came up with: In the future, we should cultivate* five kinds of minds. *Because I had so often used the word "mind" as a psychologist, I had to point out that this was not a revision of MI theory—I had not suddenly dropped three or four intelligences!*

*Rather, I was describing the* kinds of faculties and capacities that are most important in our time *and how best to cultivate them. I was addressing policymakers and the general public rather than educators—the group ordinarily charged with cultivating the mind.*

In the interconnected world in which the vast majority of human beings now live, it is not enough to state what each individual or group needs to survive on its own turf. In the long run, it is not possible for parts of the world to thrive while others remain desperately poor and deeply frustrated. Recalling the words of Benjamin Franklin, "We must indeed all hang together, or, most assuredly, we shall all hang separately." Further, the world of the future—with its ubiquitous search engines, robots, and other computational devices—will demand capacities that until now have been mere options. To meet this new world on its own terms, we should begin to cultivate these capacities now.

As your guide, I will be wearing a number of hats. As a trained psychologist with a background in cognitive science and neuroscience, I will draw repeatedly on what we know from a scientific perspective about the operation of the human mind and the human brain. But humans differ from other species in that we possess history as well as prehistory, hundreds and hundreds of diverse cultures and subcultures, and the possibility of informed, conscious choice; and so I will be drawing equally on other relevant disciplines. Because I am speculating about the directions in which our society and our planet are headed, political and economic considerations loom large. And, to repeat, I balance these scholarly perspectives with a constant reminder that a description of minds cannot escape a consideration of human values.

Enough throat-clearing. Time to bring onstage the five *dramatis personae* of this literary presentation. Each has been important historically; each figures to loom even more crucially in the future. With these "minds," as I refer to them, a person will be well equipped to deal with what is expected, as well as what cannot be anticipated; without these minds, a person will be at the mercy of forces that he or she can't understand, let alone control.

*The disciplined mind* has mastered at least one way of thinking—a distinctive mode of cognition that characterizes a specific scholarly discipline, craft, or profession. Much research confirms that it takes up to 10 years to master a discipline. The disciplined mind also knows how to work steadily over time to improve skill and understanding—in the vernacular, it is highly disciplined. Without at least one discipline under his belt, the individual is destined to march to someone else's tune. (See Essays 16–18 in *The Essential Howard Gardner on Education*.)

*The synthesizing mind* takes information from disparate sources, understands and evaluates that information objectively, and puts it together in ways that make sense to the synthesizer and also to other persons. Valuable in the past, the capacity to synthesize becomes ever more crucial as information continue to mount at dizzying rates (see Essays 35–36).

Building on discipline and synthesis, *the creating mind* breaks new ground. It puts forth new ideas, poses unfamiliar questions, conjures up fresh ways of thinking, and arrives at unexpected answers. Ultimately, these creations must find acceptance among knowledgeable consumers. By virtue of its anchoring in territory that is not yet rule-governed, the creating mind seeks to remain at least one step ahead of even the most sophisticated computers and robots (see Essays 23–24).

Recognizing that nowadays one can no longer remain within one's shell or on one's home territory, *the respectful mind* notes and welcomes differences between human individuals and between human groups, tries to understand these "others," and seeks to work effectively with them. In a world where we are all interlinked, intolerance or disrespect is no longer a viable option.

Proceeding on a level more abstract than the respectful mind, *the ethical mind* ponders the nature of one's work and the needs and desires of the society in which one lives. This mind conceptualizes how workers can serve purposes beyond self-interest and how citizens can work unselfishly to improve the lot of all. The ethical mind then acts on the basis of these analyses (see Essays 28–33).

One may reasonably ask: Why these five particular minds? Could the list be readily changed or extended?

My brief answer is this: The five minds just introduced are the kinds of minds that are particularly at a premium in the world of today and will be even more so tomorrow. They span both the cognitive spectrum and the human enterprise in that sense they are comprehensive, global. We know something about how to cultivate them.

Of course, there could be other candidates. I considered candidates ranging from the technological mind to the digital mind, the market mind to the democratic mind, the flexible mind to the emotional mind, the strategic mind to the spiritual mind. I am prepared to defend my quintet vigorously. Indeed, I've written a whole book to that effect (Gardner, 2007).

This may also be the place to forestall an understandable confusion. My chief claim to fame is my positing, some years ago, of a theory of multiple intelligences (MIs). According to MI theory, all human beings possess a number of relatively autonomous cognitive capabilities, each of which I designate as a separate intelligence. For various reasons, people differ from one another in their profiles of intelligence, and this fact harbors significant consequences for school and the workplace. When expounding on the intelligences, I was writing as a psychologist and trying to figure out how each intelligence operates within the skull.

*The five minds are different from the eight or nine human intelligences.* Rather than being distinct computational capabilities, they are better thought of as broad uses of the mind that we can cultivate at school, in the professions, or at the workplace. To be sure, the five minds make use of our several intelligences: for example, respect is impossible without the exercise of interpersonal intelligences. And so, when appropriate, I invoke MI theory. But I am speaking about policy rather than psychology. As a consequence, please think about those minds in the manner of a policymaker, rather than a psychologist. That is, my concern is to convince you of the need to cultivate these minds and illustrate the best ways to do so, rather than to delineate specific perceptual and cognitive capacities that presumptively undergird the minds.

Here they are.

## DISCIPLINED

Even as a young child, I loved putting words on paper, and I have continued to do so throughout my life. As a result, I have honed skills of planning, executing, critiquing, and teaching writing. I also work steadily to improve my writing, thus embodying the second meaning of the word *discipline*: training to perfect a skill.

My formal discipline is psychology, and it took me a decade to learn to think like a psychologist. When I encounter a controversy about the human mind or human behavior, I think immediately about how to study the issue empirically, what control groups to marshal, and how to analyze the data and revise my hypotheses (and conclusions) when necessary. Turning to management, I have many years of experience supervising teams of research assistants of various sizes, scopes, and missions—and I have the lessons and battle scars to show for it. My understanding has been enriched by observing successful and not-so-successful presidents, deans, and department chairs around the university; addressing and consulting with corporations; and studying leadership and ethics across the professions over the past 15 years.

Beyond question, both management and leadership are *disciplines*—although they can be informed by scientific studies, they are better thought of as crafts. By the same token, any professional—whether she's a lawyer, an architect, or an engineer—has to master the bodies of knowledge and the key procedures that entitle her to membership in the relevant guild. And all of us—scholars, corporate leaders, professionals—must continually hone our skills.

## SYNTHESIZING

As a student, I enjoyed reading disparate texts and learning from distinguished and distinctive lecturers; I then attempted to make sense of these sources of information, putting them together in ways that were generative, at least for me. In writing papers and preparing for tests that would be evaluated by others, I drew on this increasingly well-honed skill of synthesizing.

When I began to write articles and books, the initial ones were chiefly works of synthesis: textbooks in social psychology and developmental psychology, and, perhaps more innovatively, the first book-length examination of cognitive science. Whether one is working at a university, a law firm, or a corporation, the job of the manager calls for synthesis. The manager must consider the job to be done, the various workers on hand, their current assignments and skills, and how best to execute the current priority and move on to the next one. A good manager also looks back over what has been done in the past months and tries to anticipate how best to carry out future missions. As she begins to develop new visions, communicate them to associates, and contemplate how to realize these innovations, she invades the realms of strategic leadership and creativity within the business or profession. And, of course, synthesizing the current state of knowledge, incorporating new findings, and delineating new dilemmas is part and parcel of the work of any professional who wishes to remain current with her craft.

## CREATING

In my scholarly career, a turning point was my publication in 1983 of *Frames of Mind: The Theory of Multiple Intelligences*. At the time, I thought of this work as a synthesis of cognition from many disciplinary perspectives.

In retrospect, I have come to realize that *Frames of Mind* differed from my earlier books. I was directly challenging the consensual view of intelligence and putting forth my own iconoclastic notions, which were ripe, in turn, for vigorous critiques. Since then, my scholarly work is better described as a series of attempts to break new ground—efforts at forging knowledge about creativity, leadership, and ethics—than primarily as syntheses of existing work. (This trek is captured with reasonable accuracy by the placement of essays in this book.)

Parenthetically, I might point out that this sequence is unusual. In the sciences, younger workers are more likely to achieve creative breakthroughs, while older ones typically pen syntheses.

In general, we look to leaders, rather than to managers, for examples of creativity. The transformational leader creates a compelling narrative about the missions of her organization or polity; embodies that narrative in her own life; and is able, through persuasion and personal example, to change the thoughts, feelings, and behaviors of those whom she seeks to lead (see Essays 25–26).

And what of the role of creativity in the workaday life of the professional? Major creative breakthroughs are relatively rare in accounting or engineering, in law or medicine. Indeed, one does well to be suspicious of claims that a radically new method of accounting, bridge-building, surgery, prosecution, or generating energy has just been devised.

Increasingly, however, rewards accrue to those who fashion small but significant changes in professional practice. I would readily apply the descriptor "creative" to the individual who figures out how to audit books in a country whose laws have been changed and whose currency has been revalued three times in a year, or to the attorney who ascertains how to protect intellectual property under conditions of monetary (or political or social or technological) volatility (see Essays 23–24).

## RESPECTFUL AND ETHICAL

As I shift focus to the last two kinds of minds, a different set of analyses becomes appropriate. The first three kinds of minds deal primarily with cognitive forms; the last two deal with our relations to other human beings. One of the last two (respectful) is more concrete; the other (ethical) is more abstract. Also, the differences across career specializations become less important: We are dealing with how human beings—be they scientists, artists, managers, leaders, craftspeople, white collar, blue collar, or professionals—think and act throughout their lives. And so here I shall try to speak to and for all of us.

Turning to *respect*, whether I am (or you are) writing, researching, or managing, it is important to avoid stereotyping or caricaturing. I must try to understand other persons on their own terms, make an imaginative leap when necessary, seek to convey my trust in them, and try as far as possible to make common cause with them and to be worthy of their trust. This stance does not mean that I ignore my own beliefs, nor that I necessarily accept or pardon all that I encounter. (Respect does not entail a pass for terrorists.) But I am obliged to make the effort, and not merely to assume that what I had once believed on the basis of scattered impressions is necessarily true. Such humility may in turn engender positive responses in others.

As I use the term, *ethics* also relates to other persons, but in a *more abstract way*. In taking ethical stances, an individual tries to understand his or her role as a worker and his or her role as a citizen of a region, a nation, and the planet. In my own case, I ask: What are my obligations as a scientific researcher, a writer, a manager, a leader? If I were sitting on the other side of the table, if I occupied a different niche in society, what would I have the right to expect from those "others" who research, write, manage, lead?

And, to take an even wider perspective, what kind of a world would I like to live in if—to use philosopher John Rawls's phrase—I were cloaked in a "veil of ignorance" with respect to my ultimate position in the world? What is my responsibility in bringing such a world into being? Every reader should be able to pose, if not answer, the same set of questions with respect to his or her occupational and civic niche.

For more than a decade, I have been engaged in a large-scale study of "good work"—work that is excellent, ethical, and engaging for the participants. As detailed in Essays 28–33, I draw on those studies in my accounts of the respectful and the ethical minds.

## EDUCATION IN THE LARGE

When one speaks of cultivating certain kinds of minds, the most immediate frame of reference is that of education. In many ways, this frame is appropriate; after all, designated educators and licensed educational institutions bear the most evident burden in the identification and training of young minds.

But we must immediately expand our vision beyond standard educational institutions. In our cultures of today—and of tomorrow—parents, peers, and media play roles at least as significant as do authorized teachers and formal schools. More and more parents homeschool or rely on various extra scholastic mentors or tutors.

Moreover, if any cliché of recent years rings true, it is the acknowledgment that education must be lifelong. Those at the workplace are charged with selecting individuals who appear to possess the right kinds of knowledge, skills, minds— in my terms, they should be searching for individuals who possess disciplined, synthesizing, creating, respectful, and ethical minds. But, equally, managers and leaders, directors and deans and presidents must continue perennially to develop all five kinds of minds in themselves and—equally—in those for whom they bear responsibility.

We should be concerned with how to nurture these minds in the younger generation, those who are being educated currently to become the leaders of tomorrow. But we should be equally concerned with those in today's workplace: How best can we mobilize our skills and those of our coworkers so that all of us will remain current tomorrow and the days after tomorrow?

## REFERENCE

Gardner, H. (2007). *Five minds for the future*. Harvard Business School Press.

# Musings About a Synthesizing Mind

At the start of the third millennium, we live at a time of vast changes—changes seemingly so epochal that they may well dwarf those experienced in earlier eras. In shorthand, we can speak about these changes as entailing the power of science and technology and the inexorability of globalization. These changes call for new educational forms and processes. The minds of learners must be fashioned and stretched in five ways that have not been crucial—or not as crucial until now. How prescient were the words of Winston Churchill: "The empires of the future will be empires of the mind." We must recognize what is called for in this new world, even as we hold onto certain perennial skills and values that may be at risk.

Some years ago I read and resonated to Arlie Hochschild's 2016 book *Strangers in Their Own Land*, an account of her 5 years of sociological research in the deep South of the United States. And so, while I did not know her personally, I nonetheless wrote Hochschild a fan letter. I indicated that the book had reminded me of David Riesman et al.'s *The Lonely Crowd* (2020)—without doubt one of the most influential sociological studies of the United States.

I often write fan letters. They are sometimes answered, sometimes not, but the important thing is to bear witness—to write and post the letter and hope that it is not treated, by program or by recipient, simply as spam.

To my delight, Hochschild wrote back almost immediately. (See Figure 35.1.) She recalled, "I remember sitting as a sophomore on the front lawn of the Swarthmore campus, talking to my boyfriend—now husband of 42 years—saying, what I'd really love to do is to write a book like *The Lonely Crowd*." And now she had!

This exchange got me thinking. What Hochschild did in her book, what we've tried to do in the Good Work Project, and more recently, in our study of higher education, goes well beyond journalism. This is not meant to critique journalism—a pursuit at least as important as scholarship in academic disciplines.

But journalists have deadlines; they write according to assignment and specification; and as soon as one article has been completed and posted, the journalist, like a case-burdened lawyer, necessarily moves on to the next. Even when journalists write books, they typically work like reporters or essayists and not like academics, whose relatively glacial pace of work and fastidiousness about quotations and sources might drive journalists crazy.

## Figure 35.1. Letter from Arlie Hochschild

August 8, 2017

To: Arlie Hochschild
From: Howard Gardner

Dear Arlie,

I don't believe that we have met, though I have known about your work for many years and heard that you made a very successful visit to the school where I teach, some months ago.

Like thousands of others, I found STRANGERS IN THEIR OWN LAND a very powerful work. I've recommended it to many others, and have compared it (to those who have some sociological memory, including Nathan Glazer) to THE LONELY CROWD.

Recently I have started a blog in education. A friend and colleague, Susan Engel, wrote an interesting blog in response to the recently reported Pew finding that Republicans believe that higher education is bad for the nation. Having been immersed in your book, I decided to respond, taking on the voice of a Trump supporter from the deep South. I include a link to the blog post for your possible interest.

https://howardgardner.com/2017/08/08/republicans-are-right-college-matters/

In appreciation and with best wishes.

Howard


August 9, 2017

To: Howard Gardner
From: Arlie Hochschild

Dear Howard,

   I remember sitting as a sophomore on the front lawn of the Swarthmore campus, talking to my boyfriend, -- now husband of 42 years—saying, what I'd really love to do is write a book like The Lonely Crowd. So thank you for saying that. And I've long appreciated your illuminating work on multiple intelligences. I'm a fan back. And thanks for your thoughtful blog—we have much work to do.

           All the best,
                        Arlie

In contrast, once past formal training, those of us who are scholars typically select our own topic and spend the time (and as needed, and as available, the money or other resources) examining something systematically. We pursue the topic in as many ways as appropriate and feasible, and seek to relate it systematically to previous work in the field. We never know in advance when we will be done; and it may well be the case that at the end of the day (or, indeed, the decade), we don't find much of interest or we don't have confidence in what we think or believe that we have actually found. And so we may write it up, or not.

This combination of wandering for quite a while and then sharply focusing— *the warp and woof of synthesizing*—is what sociologist David Riesman and coauthors Nathan Glazer and Reuel Denney did in *The Lonely Crowd* and what Arlie Hochschild did in *Strangers in Their Own Land*. Shifting to psychology, it is what my teacher Erik Erikson did in *Childhood and Society*.

For sure, in these various collaborative projects, the aforementioned scholars had gathered and analyzed the data and were able to display it in various ways. But at the end of the day (or more likely, at the end of the decade), here's what's important: as scholars, *they*—and, to pivot to the personal, *we*—assembled these data and synthesized our numbers and our impressions in as powerful a way as we could. And over and above the data that we arrayed, we hoped that our writings, individually or collectively, could change the conversation about human beings at a particular time in a particular social and cultural context.

I should point out the obvious: Intent does not necessarily equal achievement. My collaborators and I have aimed to change the conversation; the other scholars I've mentioned succeeded in doing so.

Importantly, as scholars, our words, our terms, our concepts are intended to be neutral, or to use an adjective that is less familiar but more illuminating, we aspire to a disinterested stance. We attempt to describe a state of affairs as we have observed and analyzed it, not to prejudge or skew our findings and conclusions. But if we are successful, our very acts of conceptualization and writing may change how people talk and think. And, paradoxically, this very change may in the end—or, more precisely, in the next iteration—make our conceptualizations (as initially published) less accurate, less relevant, more like "period pieces" than the "last word."

A few examples:

- If we cast a spotlight on "bowling alone," perhaps people will start to bowl together. That will alter social relations in our time and, accordingly, we'll need a new way to talk about the emerging constellation of social relations.
- If we note the risks of "fast thinking," perhaps more of us will adopt slower thinking on more occasions, or maybe we'll come up with ways of oscillating among these alternative modes of thought.
- Perhaps if we describe identity crises carefully, fewer people will have them; or perhaps identity will play out differently for the "app

generation"—the title of a 2013 book I wrote with Katie Davis (see Essay 37). Indeed, that co-authored book represents a respectful critique both of David Riesman and colleagues (nowadays, we argue, young people are app-directed, not other-directed) and of Erik Erikson (identity formation may be more precocious but also less solidified in a social media world; intimacy may be more elusive; imagination may be more likely to be collective).

You may well be thinking, "This is not science," at least not science as we know it or science as we heard it described in our school days. In one sense, science also changes—today we conceptualize atoms differently than did John Dalton, and genes differently than did William Bateson. But for all intents and purposes, the atoms and genes don't change, nor do carbon molecules nor exoplanets nor geological strata (at least not quickly).

What changes, on the bases of observations, experimentation, and critiques thereof, is how we name and describe these elements and how we assemble them together into larger schemes. If science is done well, it can be done anywhere by anyone who is trained, and in the end the conclusions should be roughly the same. The state of affairs is so different in the realm of the mind, the culture, the society, that I think it is misleading to apply the same descriptor "science" to the aforementioned constructs.

## AN ASIDE

So I no longer call what I do *social science*. If I had to choose a name, I could not think of a worse one than "Social Relations" or "Soc Rel"—the field I studied in college—founded in 1946, buried in 1972, may its name rest in peace!

I'd like to sponsor a competition to come up with a better name. I actually like the phrases that I associate with social psychologist George Herbert Mead (*Mind, Self, and Society* [1934]) or with developmental psychologist Lev Vygotsky (*Mind in Society* [1978]). Interestingly, these phrases were created by editors of compilations, and not (as far as I can determine) by Mead or Vygotsky themselves.

These somewhat noun-heavy phrases capture my long-term interest—indeed, my lifelong infatuation—with the human mind, along with the realization that minds develop and change within particular societies and cultures and that those macro-environments affect them powerfully, unpredictably, ceaselessly. Another candidate phrase would be "human studies"—but that phrase risks being transmogrified into humane or humanistic studies, neither of which I have in mind. "Person" lacks that baggage, but it seems a bit individualistic, so we could pluralize it to "persons in societies."

\*     \*     \*

For now, given my 50-plus years of fascination with the human mind, I'll describe it as "human syntheses about human minds." This formulation has the advantage that it denotes syntheses done by human beings (rather than primarily by computational devices . . . at least so far), and it is about human entities and activities. But it's wordy . . . so the raffle/competition/naming contest continues.

## REFERENCES

Erikson, E. (1950). *Childhood and society*. W. W. Norton & Company.

Gardner, H. & Davis, K. (2013). *The app generation: How today's youth navigate identity, intimacy, and imagination in a digital world*. Yale University Press.

Hochschild, A. R. (2016). *Strangers in their own land: Anger and mourning on the American right*. The New Press.

Mead, G. H. (1934). *Mind, self, and society from the standpoint of a social behaviorist*. University of Chicago Press.

Riesman, D., Glazer, N., Denney, R., & Sennett, R. (2020). *The lonely crowd: A study of the changing American character*. Yale University Press.

Vygotsky, L. S. (1978). *Mind in society: The development of higher psychological processes*. Harvard University Press.

# Some Further Reflections on the Synthesizing Mind

A few decades ago, I had the pleasure of speaking to Murray Gell-Mann, Nobel laureate in physics and the founding genius behind the interdisciplinary Santa Fe Institute. Almost as an aside, Gell-Mann said, "In the twenty-first century, the most important kind of mind will be the synthesizing mind."

Not only have I never forgotten this phrase, but as early as 2005, I began to write about the "synthesizing mind": the capacity to take in a lot of information, reflect on it, and then organize it in a way that is useful to you and (if you are skilled and fortunate) that also proves useful to others.

I am often asked about which intelligences I have and which I lack. I do my best to respond reflectively. But I've concluded that it's more useful, more informative, perhaps even more accurate to say, "I have a synthesizing mind."

Accordingly, in what follows, I probe the synthesizing mind and offer my best analysis of what it is and how it works (and, along the way, what it is not). I do so in two ways:

(1) by contrasting synthesizing with the four other kinds of minds that I have described (disciplined, creative, respectful, and ethical); and

(2) by examining the act and skill of synthesizing via the lens of the multiple intelligences.

When introducing the five minds, I make two distinctions. First of all, these minds do not denote the kinds of scholarly distinctions made by psychologists about the operations of the human mind. In other words, I have not simply dropped three or four intelligences from the current list of eight or nine intelligences. Rather, they are characterizations, directed primarily at educators and policymakers, about the kinds of minds that we, and especially those of us who occupy positions as policymakers, should be cultivating in ourselves and nurturing in others in the years ahead. These are the kinds of minds we should be valuing both within and beyond formal school.

Second, two of the postulated five minds pertain to human beings living with others, both those near us (the respectful mind) and those with whom we have more remote contact (the ethical mind). My understanding of these two kinds of minds has emerged from our longtime and continuing study of good work, good play, good citizenship, and other kinds of "goods." These minds are incredibly important for the survival as well as thriving of our species, but they are not relevant to the analytic task that I have set myself here.

The remaining three minds describe thinkers, learners, teachers—the minds of cognition. Let me introduce them briefly.

In conceptualizing the *disciplined mind*, I emphasize what it takes to master the major ways of thinking that thinkers (often scholars) have developed over the centuries: the minds of the philosopher, the psychologist, the economist, the physicist, the historian, the musical composer, the musicologist, and so on, right through the college course catalogue. We may differ from one another in our potentials to "do" biology or biography or ballet, but in any event, it takes years to master these and other disciplines. I contend as well that meaningful synthesis or meaningful creativity are not possible until or unless one has developed a mind that has mastered one or more disciplines. To play with words, it takes discipline to master a discipline. Rarely are there individuals who synthesize skillfully without formal training in one or more disciplines.

Let me now introduce—or reintroduce—the other two cognitive minds (see Essay 34). In its pristine form, the *synthesizing mind* takes as its assignment the optimal organization of materials from one or more disciplines or fields or spheres of observation and tinkering. The materials should be organized in ways that are accurate, comprehensive, and suited for the task at hand. Many assignments in school—for example, book reports, term papers, essay questions on an examination—call at least modestly for the synthesizing mind, and indeed, that is what textbooks at their best should present and what most courses should engender in dutiful students.

At its core, the *creating mind* solves problems or raises questions or introduces ideas or practices that are initially novel, if not unprecedented. But novelty alone is not enough—just thinking or doing "stuff" in an original way does not suffice. To be judged creative, an idea or practice must be accepted in some way by a relevant community (see Essays 23–24). Put whimsically (or perhaps wistfully), it's not enough that your mother or father posts your finger-painting on the refrigerator door. The scribble or sketch has to capture the eye of the curator or the art critic or the writer of a textbook and/or the pocketbook of the collector, and eventually emerge as worthy of attention by other members of the public. And occasionally, the reverse process happens: For some reason, a run-of-the-mill action or product somehow goes viral and then the creator, critic, or collector takes note.

Now we come to the heart of the matter: the line between synthesizing and creating. It's hard to imagine any potentially creative idea or act being conjured up "from scratch." The 10 years of disciplined training really matter. Even such indisputable prodigies as Mozart or Picasso in the arts, or indeed Bill Gates or Steve Jobs in computing, proceeded in workmanlike fashion in their chosen domain for years before achieving a creative breakthrough. So creativity presupposes both some disciplinary mastery and some prior synthesis.

But it is worth drawing a line—even a sharp line—between syntheses that are content to be syntheses and are accepted as such, and syntheses that aim for and may be accepted as creative by the relevant community.

To use my own quite humble example: A textbook in social psychology that I co-authored in the late 1960s (Grossack & Gardner, 1970) is a mundane synthesis of major topics in the field and has had no appreciable impact on how social psychology is conceptualized or presented in subsequent texts.

In contrast, my textbook in developmental psychology (1978) is more innovative. It is organized by developmental stage rather than by topic (language development, social development, and so on); and it features interludes—short essays interspersed between chapters where I reflect on topics that happened to interest me (and, I hoped, students), rather than on topics already addressed by the research community or already on the radar screens of classroom teachers. And ultimately this textbook has had some impact on future texts in that field and, one would hope, on the ways in which students and teachers think about human development.

Unlike my earlier text in social psychology, the developmental text gets at least modest marks for creativity.

Looking over my oeuvres of a lifetime, it is clear to me that most of my early works are primarily syntheses; two textbooks, my study of Piaget's and Lévi-Strauss's *Quest for Mind* (Gardner, 1973), my survey of *The Shattered Mind* (Gardner, 1975), and so on. I was displaying what the good and conscientious student has mastered. At its inception, the work on multiple intelligences was seen as a synthesis. In fact, as previously noted, it began with my initial intention of surveying different kinds of minds. But *Frames of Mind* (Gardner, 1983) moves significantly toward the pole of creativity in several ways:

- it considered a far wider range of abilities than those usually considered by psychologists interested in cognition and intelligence;
- it surveyed a far wider range of scholarly disciplines—from cultural studies to brain studies—than was typically the case;
- it blurred the line between natural and social scientific research findings, on the one hand, and educational implications on the other; and, perhaps most prominently, if less predictably,
- it pluralized and conferred new denotations and connotations on the venerable word "intelligence."

Quite possibly more than anything else that I'd written before or since, *Frames of Mind* is a *synthesis that has significantly changed the conversation about human intellect*, particularly with reference to educational policy and practices. And more evocatively, it has catalyzed many individuals to think differently about themselves and about others—often persons to whom these individuals were close but whose minds had previously been mysterious, even opaque. And so it deserves at least a modest nod in the direction of creativity.

Syntheses can be adequate or inadequate, suited for one purpose or for a quite different one. But no matter the ambition or modesty of the synthesizer, he or she cannot simply decide or declare that a specific synthesis is also creative. That's a decision made only and appropriately by the wider community over the course of time.

And so, as an example, Katie Davis and I hoped that young people in the 21st century would, following the examples of David Riesman and Arlie Hochschild, come to be thought of as "the app generation." But that change in conceptualization and terminology has not happened yet, and it probably never will. Alas, it fails the acid test for creativity.

A few more points about synthesizing. Some syntheses aim to bring lots of materials together to make a single grand point. We could call those "hedgehog" syntheses. Charles Darwin's *On the Origin of Species* is an outstanding example.

"Fox-like" syntheses delight in their plurality—Carl Jung's positing of various personality types might qualify.

Some syntheses seek *balance*—Riesman et al.'s (1956) and Hochschild's (2016) portrayals of American life come to mind—while others push very hard in one direction or other—C. Wright Mills (1956) on "the power elite" and William Whyte (1956) on the "organization man" were both attempts to characterize American society in the 1950s.

And within the area of social analysis, some syntheses reflect *great overarching concepts*—the "grand theories" of Karl Marx, Max Weber, Sigmund Freud—whereas others are content to focus more deeply on a more manageable topic, what sociologist Robert Merton termed "theories of the middle range" (1968).

But by now, you may be posing the question that is most challenging for me to answer: Is there a separate intelligence for synthesizing, or is the capacity that I am describing adequately explained by the operation of one or more of the several multiple intelligences?

Of course, everyone is free to claim or create the category of "synthesizing intelligence"; and I willingly concede that I sometimes have that bent or at least that aspiration. And I can readily cite authors who are expert synthesizers; let me mention paleontologist Stephen Jay Gould and geographer Jared Diamond, both of whom are also steeped in history and, as it happens, in music.

Without stretching the point too far, I can talk of artists who also synthesize styles and formats. For example, after widely heralded creative breakthroughs in the opening decades of the 20th century, composer Igor Stravinsky and his contemporary (and occasional collaborator) visual artist Pablo Picasso both went through a neoclassical period in the 1920s. As the name signals, both artists combined contemporary idioms with classical themes and forms.

And some artists even deliberately synthesize media, such as opera composer Richard Wagner, with his *Gesamtkunstwerk*. (This term denotes works that draw on several art forms in an effort to present a grand unified vision [or versions] of the world. It's also an apt characterization of artists who attempt to capture an entire universe. Consider the scope of the Inferno, Purgatory, and Paradise of Dante's *Divine Comedy*.)

In contrast, some scholars, musicians, and painters are equally notable but explore the same topics and themes ever more deeply, rather than more widely or broadly. I think of a cognitive psychologist who has studied metaphoric and analogic thinking for 40 years, or a neurobiologist who has focused for an equal length of time on a set of cells in the retina.

Still, as noted in my review of MI theory (see Essays 17–20), I am very conservative with respect to conferring a seal of approval on any additional intelligences. I much prefer to construe the candidate skill in question as one that deploys two or more already approved intelligences in the cohort. For example, I have never been tempted to posit separate "technological" or "digital" intelligences. Such candidates are explained to my satisfaction by combinations of the already identified intelligences.

In a nutshell, here's my idea: Individuals can synthesize in various ways. How they do so depends on the intelligences that most characterize their own cognitive profile—the ones on which they most like to draw, and the ones most appropriate for the task at hand. Quite obviously, a composer like Stravinsky draws heavily on musical intelligence, while a painter like Picasso draws on his spatial and bodily intelligences. Such creative artists may well have strong linguistic and logical intelligences, but those "IQ" strengths are bonuses rather than requirements. As it happens, Stravinsky, trained as a lawyer, wrote prose that was sparkling and illuminating in several languages; in contrast, Picasso was a painting prodigy, and his writings and sayings are novelties, nothing more.

Synthesizers like the aforementioned Gould or Diamond naturally draw on the intelligences that scholars use, and so both of them presumably have abundant linguistic and logical intelligences. But they probably have more developed linguistic intelligences than, say, Murray Gell-Mann, the unwitting coiner of the term "synthesizing mind," who signed a contract to write a book but had enormous difficulty in completing it.

In contrast, many brilliant physicists, mathematicians, and computer scientists would rather tinker than talk, never have any urge to write out what they have been thinking in more than a few pages, or prefer mathematical symbols to linguistic tropes. I began to write regularly, if not compulsively, as soon as I could hold a pencil or type on a keyboard or place letters on the plate of a printing press; over the decades I have rarely searched for numerical signs or symbols, let alone felt any urge to create new ones.

So far my descriptions, though admittedly based simply on biographical information and not on "tests," seem plausible. It would be hard, if not perverse, to state that dancer Martha Graham lacked bodily intelligence or that physicist Marie Curie lacked logical-mathematical intelligence. Indeed, if psychological tests were to challenge those characterizations, I would question the validity of those tests!

But I suspect that those of us who are synthesizers also draw on our other less obvious intelligences in ways that prove helpful to us and perhaps to others. For example, as a physicist, Einstein would probably be seen as a prototypical master of logical-mathematical intelligence. Yet he actually had a highly developed spatial intelligence, while he needed the help of his friend Marcel Grossmann to execute the mathematics appropriate for the general (as compared to the special) theory of relativity. As a scientist and physician, Freud had adequate logical and spatial intelligence, but he really stood out for his linguistic skills; in 1930 he won the prestigious Goethe Prize for literary excellence.

Since I am indulging in the just-contrived parlor game of analyzing synthesizers in terms of their most helpful intelligences ("name the intelligences of synthesizing"), let me peer into my own MI mirror. As I've reflected, my linguistic intelligence is fine, and my logical skills are certainly adequate. As for interpersonal intelligence, this should be a requirement for psychologists, at least those who study or seek to help human beings.

But as I consider my own approach to synthesizing, I have come up with a somewhat different analysis. As a synthesizer, I believe that I draw readily on two other intelligences.

First of all, as one deeply involved with classical music since early childhood, I think of writing, and particularly the writing of a book, as if it were the creation of a symphonic composition, with movements, themes, anticipations, recapitulations, interludes, and the like. Both with respect to my own writing and that of students and colleagues, from early on, I have an overall sense of the form and structure of the final piece, and in my case I like to pound out a detailed outline or even a whole draft early on. I have a developed sense of the order in which themes should be introduced; when various "instruments" should be foregrounded; the implications of moving a text or a point to a different spot in the literary tapestry, or of expanding or eliminating it; what serves as an overture, the development of a theme, a climax and conclusion, even a coda (stay tuned!).

Perhaps I am fooling myself that I do this effectively, but I believe that I *do* do it, and I strive to do it reasonably well.

The other perhaps surprising intelligence on which I draw is that of the naturalist. In truth, I don't have much interest in the natural outdoors, despite spending all those years at summer camp and passing all of the relevant merit badges needed to become an Eagle Scout. But as I try to make sense of all that I have seen, heard, read, and thought, I am constantly coming up with schemes, ways of classifying and reclassifying, tables, simple images, orderings and reorderings.

To be sure, this is hardly in the league of naturalists like Linnaeus or Darwin or Audubon; but it represents a legitimate effort to make optimal sense of the variety of "species" that I have collected. And that may be why I have always been far more attracted to biology than to the other physical or natural sciences, and why I enjoyed my years working as a neurologist "from the neck up." Still, I suspect I would have been a more successful biologist in the 19th or early 20th century than in the current era, where mathematics and computing are increasingly important, and classifying is more likely to be done by a computer program or deep-learning algorithm than by a human eye and a human brain.

Of course, all of this is self-analysis. And if I can poke fun at my arsenal of intelligences, the truth of it depends on whether I have keen intrapersonal intelligence. But as I think about synthesizing more broadly, beyond my own perhaps peculiar personal psychology, I propose a few additional points.

To begin with, synthesizing involves a *touch of artistry*. To create a work or treatment of significant size and to convey it effectively to others requires a sense of form, of arrangement of parts, of the attention and predilection of audiences, of launching, development, and closure. All of these skills are identified particularly with the arts more so than with the sciences, where the behavior and accurate description of reality has to take primacy. I propose that synthesizers are aspiring artists, and that they—and that includes me—draw from the art form with which they are most comfortable—be it musical, literary, architectural, terpsichorean, or some other aesthetic means of communication.

But unlike "pure" artists, synthesizers cannot start from scratch and proceed in any and all directions. Rather, they are restricted, or, if you prefer, empowered, by *the data that they and others have collected*, be it historical, literary, or psychological. And accordingly, these aspiring synthesizers need some ways to arrange and rearrange those data, until they find a solution that is adequate, accurate, communicable, and, as a bonus, aesthetically pleasing. In my case, I proceed in the ways of a naturalist, labeling and relabeling different species of data. But I suspect that there are many ways of ordering and reordering—perhaps even as many as there are synthesizers or at least types of synthesizers.

I am prepared to go further. Just as individuals draw on their favored intelligences in mastering schoolwork or in creating new works—so, too, those of us who are engaged in synthesizing are inclined to exploit those intelligences that we favor, and that help us to make sense of our experiences. Think of the intelligences as a chart of chemical elements: Aspiring synthesizers create the particular configuration of chemical compounds that allow them to carry out their mission.

Psychology has largely dropped the ball with respect to illumination of this form of cognition. We have little understanding of how we humans synthesize information and then collate it in ways that are helpful to us and to others, and in ways that are either adequate, fine, outstanding, or truly creative. This skill does not lend itself to examination via simple laboratory experiments and concomitant rapid publication in highly ranked peer-reviewed journals; hence the paucity of relevant research on the kind of broad synthesizing that I am describing. This state of affairs constitutes an inviting challenge for students who enjoy synthesizing, and who may be able to illuminate its operations, at least until such time as a master "deep learning" algorithm renders anachronistic the synthesizing intelligences as practiced by human beings.

My own case study here provides possible starting points for such an enterprise. A future student or scholar of synthesis might carry out comparisons at a certain moment in time: the rise, breadth, and reasons for curiosity in early childhood (for example, what it means to grow up in a household where important information has been concealed and where so much of the ambient society is unfamiliar to the adults who grew up elsewhere); the ways in which a motley collection of information is stored and retained in the absence of intellectual coatracks; the extent to which those packets of information are linked or allowed to float in cognitive space; the impulse to record tentative syntheses in an appropriate symbolic form; the urge to put together disparate bodies and forms of information; the tendency to master, dabble in, challenge, or circumvent existing disciplines; the stretch (or nonstretch) to creativity (for example, whether one is talking about classroom assignments or dissertations, or the production of reliable summaries as opposed to the search for ambitious breakthroughs); as well as the difficult decision of what to leave out, or to postpone for another day or another publication.

Or, alternatively, such a student of synthesis might carry out longitudinal case studies of individuals with clear synthesizing gifts and compare them with one another, or, indeed, with equally talented individuals who are perhaps outstanding disciplinarians or creators, but display no particular inclinations or gifts in synthesizing or whose syntheses are decidedly modest, misleading, or even useless. (I've fantasized about a study of four individuals nicknamed Steve—geologist Gould, physicist Weinberg, psychologist Pinker, literary scholar Greenblatt—and how each of them has become a master of synthesis. Or what about anthropologist Margaret Mead or Mary Douglas, novelist Margaret Drabble or Mary McCarthy, and historian Margaret Macmillan or Mary Beard?)

Then there is the question of "educating for synthesizing." As already noted, we make modest requests for synthesizing when we ask students to write book reports or carry out a project or produce an essay in response to a probe on a test. There are criteria for judging these productions (ones that our students push for if we teachers don't lay out the rules for rendering such judgments, preferably in excruciating detail). Schoolwork beyond these modest requirements is generally a term paper or a thesis in a discipline, where the student is expected in some way to stretch beyond the existing literature or already secured experimental results. If, indeed, the synthesizing mind is important in our time, it seems well worth efforts to develop it systematically and skillfully.

In deciphering how to navigate large bodies of knowledge, especially knowledge drawn from disparate disciplines, and how to do so in a way that is effective and that moves the conversation along, even if it does not change it, scholars are largely on their own. There are remarkably few guides. And so, as in my case, aspiring synthesizers have to depend on role models—those known personally and those one has learned about. I have been fortunate in that regard. We aspiring synthesizers must draw on our peculiar motivations and skills and intelligences, such as they are and such as they can be cultivated or even synthesized.

But in the future, there's no need for budding synthesizers to proceed on their own. Even to introduce the "word" and the "concept" of "synthesizing" in school would be a positive step. Instructors can model how they go about making sense of disparate data, as I have often done with my students, both in a classroom setting and in one-on-one conversations with doctoral students. One can study the syntheses that we admire and that are useful to us, as well as those that are confusing or that leave us uninspired (comparing how different textbooks cover the same topic can be enlightening).

And most important, one can explicitly assign "challenges that require synthesizing" to students: ask for their planned approach, provide feedback on these "specs," look at drafts and provide further feedback, then have students read or examine the efforts to their classmates and indicate where they inspire and where they fall short. Indeed, such procedures are often followed with respect to lab reports or book reports or, more ambitiously, to term papers or theses—what I'm calling for is a much more explicit concern with what makes for an effective synthesis and the various ways in which such synthesis can fall wide of the mark.

Synthesis occurs in many domains of practice. Many artists carry out synthesizing in the symbol systems favored by their art form. Indeed, synthesizing occurs across many domains of practice. Think of the judge who has to make a decision about a complex case and then issue a decision; the physician who has to cull all sorts of tests and observations and come up with the correct diagnosis; the curator or the city planner who has to mount an exhibition or a celebratory event; the corporate leader who decides to create a new business plan; or the management consultant charged with providing useful feedback to an enterprise that has foundered. One could write a book about "varieties of syntheses" and "varieties of synthesizers."

A final word about synthesizing. The kinds of synthesizing on which I have focused might be called large-scale synthesizing, the sort that scholars undertake and that is found in books, summary articles, or major artistic or design endeavors. But we also live in an era where rewards come to those who can synthesize in a versatile way: provide a brief summary of a complex set of ideas; write a blog; give a TED talk; do an effective interview on radio, TV, or a podcast; create a stimulating tweet and marvel if it goes viral.

But—an important "but"—can such a quick and necessarily superficial synthesizer go deeper or deep, provide more details, handle challenges, or realize when a criticism is valid as opposed to being irrelevant or based on a fundamental misunderstanding? Only if the answer to these questions is "yes" would I deem such a person to be a legitimate synthesizer.

Murray Gell-Mann spoke about the synthesizing mind as central in our century, and I believe that he was correct. But this claim raises the issue of whether it will continue to be important thereafter, and if so, whether the synthesizing will be done by human beings, by artificial intelligence, or by some combination of protoplasmic and "chip" machinery.

I lack expertise in this area. To be sure, I have no doubt that aspiring synthesizers will appreciate and make use of the most powerful and most appropriate computational devices and resources. As a modest example, in the Good Work

Project in the 1990s, we did most of our analyses by reading and coding and applying a few statistical tests. In contrast, in our study of higher education two decades later, we have been using "big data" programs to analyze the words and phrases used by 2,000 constituents across several campuses.

But computational approaches will also stretch us in unanticipated directions. As just one example, most individuals today are trained in specific disciplines or forms of expertise and draw on the tools of their respective disciplines. In contrast, deep-learning algorithms simply process data without such classificatory bins, unless they are so "primed." Such "undisciplined" minds may reveal patterns and interactions that would otherwise have been missed. Someone might even create a "Howard Gardner synthesizing app" that synthesizes in the way that I would have, but does a better job than I could have. In which case I *really* could retire!

Personally, I am unenthusiastic about allowing computational devices to determine which questions should be asked, to dictate what should be synthesized and how, to judge whether the particular synthesis is appropriate for the task and the user, and to decide how the results should be interpreted. Nor am I by any means ready to turn over moral or ethical decisions to an algorithm, no matter how intelligent it is deemed or declared to be. However, there are likely to be goals and processes of synthesizing à la AI that are difficult even to envision at this point, at least by me.

From early on I have been disposed to ask questions, to reflect about them, to gather whatever data are available by whatever means I have at my disposal, and then, crucially, to organize those data in ways that make sense to me, and, I hope, to others as well. Thanks to the accidents of family, culture, the historical moment, and, yes, genes, I have had the privilege of being able to lead a life of the mind, and to put forth syntheses that make sense to me, and, at least at times, to others as well. I have done this chiefly through writing books, though as other media have become available or salient, I've been able to draw on them as well. And if I continue to be fortunate, I'll be able to continue to do this for a while, and to support others—students near and mentees far and correspondents wherever—who seek to do the same.

Indeed, there is an apparent advantage to the synthesizing impulse, which I certainly had not anticipated when I was younger. As one ages, and one's memory for names or appointments or the location of car keys declines, the impulse and even the capacity for synthesizing seems to remain intact. One might even call it wisdom.

Regarding lessons for others, a few closing comments.

First of all, should you have the privilege of changing the conversation, be grateful. Don't assume that you can control the ensuing conversation—you will in all likelihood fail—but you do have a responsibility to help guide it in productive ways. And if synthesizing in the area of human minds and human activities is appealing, go for it. Perhaps you will even be able to explain how the synthesizing mind develops and how it operates in full throttle.

Second, try to explain your enterprise to others in the most suitable terms and concepts. For too long, I thought that I had to justify my work in terms of "harder" disciplines, be it logic and mathematics on the one hand, or biology, neurology, and genetics on the other. I don't think that such strained rationales are necessary or wise. Rather, try as best you can to describe just what you have done, with what evidence and analytic tools you have, while indicating as well what you have not done and what you are not able to do.

And don't assume that others will know what you are doing or why you have done it, unless you can explain it clearly and, if necessary, repeatedly.

In my case, I have not done journalism, nor have I done science with a capital S. Focusing on human cognition in human societies, I have sought to carry out syntheses that are useful; to indicate how I have done them; and, if fortunate, to affect or even change the conversation, and to be open to that change bringing about yet other changes, which we might not have anticipated but that we can, if we so choose, then critique, evaluate, and build on.

The cycle continues.

May the human conversation—better, the human conversations—also continue. May synthesizing minds thrive!

### REFERENCES

Gardner, H. (1973). *The quest for mind: Jean Piaget, Claude Lévi-Strauss, and the structuralist movement.* Knopf.
Gardner, H. (1975). *The shattered mind.* Knopf.
Gardner, H. (1978). *Developmental psychology: An introduction.* Little Brown, International Edition.
Gardner, H. (1983). *Frames of mind: The theory of multiple intelligences.* Basic Books.
Grossack, M., & Gardner, H. (1970). *Man and men: Social psychology as social science.* International Textbook.
Hochschild, A. R. (2016). *Strangers in their own land: Anger and mourning on the American right.* The New Press.
Merton, R. K. (1968). *Social theory and social structure.* Free Press.
Mills, C. W. (1956). *The power elite.* Oxford University Press.
Riesman, D., Glazer, N., & Denney, R. (1950). *The lonely crowd: A study of the changing American character.* Yale University Press.
Whyte, W. H. (1956). *The organization man.* Simon & Schuster.

# Why the App Generation?

*Howard Gardner and Katie Davis*

*Of course I have been aware of the computer revolution since my student days, and wrote about artificial intelligence in my 1985 study of cognitive science (see Essay 21).*

*That said, most of my research has not been focused on computers, digital media, and artificial intelligences. It's largely because of opportunities to collaborate that I've focused more on that sphere/slice/stance of mind.*

*I am especially indebted to the MacArthur Foundation. As a member of a research group on "Youth and Participatory Politics," I had the opportunity—with colleagues—to examine a number of issues that are central in the new digital media, and how they affect education and personal development.*

*Accordingly, I note here three specific lines of work.*

*1. With Emily Weinstein, I have revisited issues of creativity—and we have posited a fourth kind of creativity that is focused particularly on the affordances of social media (see Essay 24).*

*2. Wearing my hat as a student of the professions, I consider the challenges posed to the professions by digital media and the need—indeed, the imperative—for professionals to maintain a stance of disinterestedness (see Essay 32).*

*3. With Katie Davis, I have studied what we termed "the app generation." These are individuals, born in or shortly before the 21st century, whose lives are dominated by smart devices that are at their beck and call. We make a distinction between the positive uses of apps (which we call enabling) and those that are limiting (which we call dependence)—and, of course, we hope that future youth can use apps in a constructive and imaginative way.*

In 1950, with two younger colleagues, lawyer-turned-sociologist David Riesman published *The Lonely Crowd*. This portrait of changing American values had the rare distinction of being both a bestseller at the time *and* a classic of scholarly writing, quite possibly, second only to de Tocqueville's *Democracy in America* in its influence over the decades.

Though not quite of Twitter brevity, the argument put forth in Riesman's book can be easily summarized. In the 18th century, American colonists were *tradition-directed*: they looked to those who came before, typically from Europe, for patterns of what to believe and how to behave. In the 19th century, the new Americans were *inner-directed*. Having broken the shackles of European mores, often setting out to the far frontier, they sought to develop their own internal compass. By the middle of the 20th century, Americans had become *other-directed*. Living primarily in cities and suburbs, they consumed the mass media and sought to buy, behave, and believe in the ways that their neighbors did.

Some 60 years later, it is certainly possible to discern vestiges of tradition-directed and of inner-directed Americans. And the enduring influence of peer behavior and values underscores the prescience of the Riesman analysis; we still want to keep up with the Joneses, and increasingly with the Jals, Jins, and Juans from other cultures.

Yet, as Katie Davis and I have argued, the behavior of Americans, and especially of young Americans, is increasingly directed not by other human beings, past or present, but rather by *apps*—technologies that are held in the hand, activated by a touch of the finger, and so constituted that they carry out more and more of the tasks and asks of daily life. Indeed, in a tribute to Riesman and colleagues, we speak of this as an "app-directed generation."

Born after 1990, members of this generation have been surrounded and immersed to an unprecedented degree by technology: laptops, pads, smartphones, handheld devices of every size and shape. Unlike so-called digital immigrants, they cannot remember a less technological era; indeed, unlike every previous generation, most of them do not even know what it is like to be lost, for if they've lost their handhelds, their parents still have theirs! Asked to list the most important events in their lives, they are the first generation not to mention political, economic, or cultural events; instead they talk about the hegemony of the Internet, the web, and an ever-evolving spate of social media.

But it is not just technology in the broader sense; the particular contours of apps are at a premium. This generation expects that, like an app, *every aspect of life* will be quick, efficient, streamlined, available immediately on demand, and tell you what to do, how to do it, how others feel about it, and, at least implicitly, how you should feel about it and how you should feel about yourselves.

## ENTER APPS

Applications are streamlined shortcuts. They allow us to do what we want to do quickly and efficiently, with no wasted motion. Since the advent of computers, and especially the birth of personal computers, such applications have existed in effect. And with the arrival of smartphones and pads, many of our lives, and nearly all of the lives of our children, are punctuated throughout the day by the activation of one app after another.

Apps, of course, are a form of technology; they are neither good nor bad in themselves. But an "app mentality" or "app consciousness," the search for an app for everything, the almost automatic activation of an app, and a concomitant reluctance to proceed if an app is lacking or inefficient or deceptive can be very limiting. And it is especially pernicious in the event that this mentality becomes the dominant mode, the dominant metaphor, for how one leads one's life.

Just that, we believe, has happened to many of our young people, and to those who have their backs. Not only do they use one app after another, like turning pages in a book, but they begin to think of their lives, and the lives of others, as an ineluctable series of apps, the imagined sequence of steps introduced at the beginning of this essay. We have termed this "the super-app of life."

Of course, if life really proceeded like a prepackaged app, things could be fine. After all, it's easier to follow a script than to write a new one. But of course, no life can be completely planned; there are always accidents, unexpected losses and detours, disappointments and reorientations. And if you expect everything to proceed in lockstep fashion, then you will be completely unprepared for the twists and turns that are part of the human condition; those zigzags that, whatever the associated pains, are what make life worth living, indeed what constitute a life.

App-directedness affects every sphere of life. One's identity, as presented on one's home page on Facebook, should be clean, neat, easy to capture, and positively rendered, with no time or space for the exploration of alternative or complex identities. One's relation to others should be easily described, efficiently executed, easily joined and disbanded with equal ease; difficult face-to-face encounters should be avoided as much as possible, and deep probing of motivations should be shunned. Indeed, so appealing is the app approach that many individuals come to think of their lives as one app after another: the right secondary school, the right summer camp, the right college, the right extracurriculars, the right internship, the right job (in Silicon Valley or on Wall Street, of course), and a great chain of being spanning cradle to crematorium. We call this a super-app.

Strikingly, people today have diametrically opposed views of this phenomenon. For some, a world of apps approaches utopia; how wonderful it is to be able to accomplish whatever one wishes readily and then move on to the next assignment or enticement! For others, an app-suffused world is a dystopia; all aspects of human agency, individuality, and creativity have been reduced to an algorithmic formula.

We take a less apocalyptic stance. In our view, apps will neither create nor destroy utopias. We prefer to contrast a life of *app enablement* with a life of *app dependence*.

The *app-dependent person* relies completely on the existing set of apps; and if an app does not exist, or is not sufficiently agile, he simply abandons course. An *app-enabled* person makes ample use of apps. But rather than simply following the app blindly, he allows the app to free his time and energy so he can take in more of the world, have richer experiences alone or with others, and treat app options as hints rather than dictates.

We like to reflect on the life path taken by Steve Jobs. Clearly Jobs admired apps, and he probably had as much to do with their invention and proliferation as any person on the planet. Yet just as clearly, Jobs never slavishly followed anyone else's script. Indeed, in his life, one can see signs of tradition directedness (his interest in Eastern spirituality), inner-directedness (his following his bliss through lengthy *Wanderjahre*), and other-directedness (his keen attention to what rival media and technology companies were doing). And perhaps that is the best lesson from a Riesman analysis: We should try to combine the strengths of tradition, our own strivings, the examples of others, and the affordances of technology.

## REFERENCES

Riesman, D., Glazer, N., & Denney, R. (1950). *The lonely crowd: A study of the changing American character.* Yale University Press.

Tocqueville, A. D. (1835). *Democracy in America.* Signet Classic.

# MISCELLANEA

I believe that the preceding essays cover the major areas of scholarship in which I have worked and also convey the order in which they were undertaken. But occasionally I have written pieces that, while not fitting neatly into that schema, seem worth preserving. And so I close this volume with two such essays.

# Human Potential

## A Forty-Year Saga

When I was growing up in Scranton, Pennsylvania in the 1940s and 1950s, I never thought of myself as a future scholar. (Yes, President Joe Biden was born in Scranton, and we are just about the same age!) In fact, as the prototypical Jewish boy (the son of immigrants from Germany who arrived in the United States "just in time"), I assumed that I would become a doctor or a lawyer—and most of our family friends had me pegged as a future lawyer.

It was only when I attended Harvard College and found that I identified with my teachers in the social sciences that I begin to consider a career as an academic. And as I relate in a memoir (Gardner, 2020), I enjoyed the life of the mind more than the pursuit of a particular discipline. I might equally well have become a historian, a biologist, a musicologist, or even a law or medical school professor. But in the end, I chose the qualitative social sciences, with a focus on developmental psychology.

Scholars typically begin their work by building on the contributions of their own teachers as well as on the achievements of those whose works they have studied. I was no exception—I began by studying the works of Jean Piaget in developmental psychology. And when I began to carry out empirical studies, I used Piagetian methods to study the artistic development of children, which Piaget had neglected because of his focus on logical and scientific thinking.

Through a series of plans and accidents, I conducted research at two different sites: Harvard Project Zero, where I was studying the development of cognitive and symbolic skills in children, particularly those employed in the arts; and the Veterans Administration hospital in Boston, where I studied the breakdown of symbolic and artistic capacities in individuals with acquired brain damage. Being (for whatever reason) more of a book-writing than an article-writing scholar, I authored books on artistic development in children, on Piaget's theory, and on the breakdown of cognitive capacities under conditions of brain damage.

Had I not gone through my old files from a few years ago, I would have forgotten that in 1976 I had outlined a book called *Kinds of Minds*. In that never-to-be-written book (and there are a few of those in my basement), I planned to describe the kinds of cognitive capacities that I'd seen developing in children and the ways in which those cognitive capacities break down under conditions of brain damage. As the grandiose title signaled, I was prepared to argue that human beings can foreground different kinds of minds—for example, the scientific mind, the artistic mind, the mechanical mind, and so on.

I put that project aside, at least for a while. Then, in 1978–1979, I had one of those experiences that end up being life-changing. A Dutch foundation, the Bernard van Leer Foundation, approached the Harvard Graduate School of Education (HGSE), where I was a non-faculty researcher, living on (or off) "soft money." The Foundation was prepared to offer the school a very generous grant—ultimately well over $1 million over a 5-year period—to answer a broad and amorphous question: "What is known about the nature and the development of human potential?"

The dean of HGSE, who happened to serve on the board of the Foundation, was also looking for ways to cover some of my salary (and that of one other junior member of the faculty). And so he asked whether, with the support of a few key senior faculty members, we would be willing to lead a "Project on Human Potential." At the time, with an eye toward Berkeley, Palo Alto, Malibu, and Esalen, I quipped, "Human potential is more of a West Coast term than an East Coast term," and yet I was quite happy to accept a leadership position and to devote some years of my life to exploring this wide-open and mind-opening question.

"Human potential" turned out to be somewhat of an inkblot test. Philosopher Israel Scheffler probes the meaning of the term in a thoughtful book called *Of Human Potential* (1986). Anthropologist Robert LeVine and sociologist Merry White looked at human potential as it is conceptualized and realized across a variety of cultures and cultural settings; their conceptualization and conclusions are reported in *Human Conditions: The Cultural Basis of Educational Development* (1986).

In my case, generous support from the Foundation allowed me, aided by several excellent researchers, to carry out a far-ranging examination of the social scientific and natural scientific evidence about various human cognitive capacities—as it were, the research needed to lay out and support a systematic argument about "kinds of minds." To embrace a term that I have subsequently adopted, I applied my "synthesizing mind" to these diverse concepts and data.

Very briefly, in *Frames of Mind* (1983), I developed the idea that intellect should be conceptualized as pluralistic. Rather than thinking of individuals as "smart" or "dumb" across the board, one can accrue considerable evidence, from a range of disciplines, that intellect is better conceived as consisting of a number of relatively independent computational devices; I elected to call these "multiple intelligences" (which I soon abbreviated as "MI").

Of all of my scholarly work, MI had by far the most immediate and, at least so far, the greatest long-term impact. While relatively few psychologists have embraced the theory (due to their allegiance to the concept of general intelligence, abbreviated as $g$), my formulation has seemed plausible to many biologically oriented scholars. And it has attained and maintained popularity within education and with the general educated public, both in the United States and abroad.

While researchers have rarely followed up my work directly, I think it has held up pretty well in the intervening decades. (And I might add that there have been hundreds, if not thousands of doctoral dissertations on "MI.")

From the vantage point of several decades, I have a reasonably clear idea of how the view of human potential captured in my writings of the 1980s represented a scholarly advance, as well as the ways in which it was limited.

In comparison to other scholarship on intelligence, I drew on a far wider set of disciplines (neurology, genetics, anthropology) and had a much more capacious view of intellect—not only solving school-style problems, but also creating products that are valued in one or more cultural settings.

I was not tied to a particular kind of test in a formal setting—indeed, I was strongly biased toward ordinary (and on occasion extraordinary) behaviors in natural settings. I was much more open to the kinds of abilities that were valued in prehistoric times (hunting, gathering, fishing, farming, divining) and to the ways in which scholastic settings have also called on different abilities in different settings at different historical periods.

Of course, my examples and evidence still drew heavily on Western research in recent decades. But when I was criticized for cultural bias, I responded, "Well, I am sure that my work is influenced by my own background and the society in which I have lived, but it is far less biased than work on intelligence undertaken by most other scholars."

But today, with the advantage of hindsight, access to new research, and considerable knowledge of what has happened in the world since the late 1970s, I would designate several limitations. And these limitations, in turn, suggest how, looking forward, one might formulate "human potential" more capaciously.

1. While I was trained across the social sciences (in a field that encompassed sociology and anthropology), I was thinking and writing very much as a psychologist. And while psychologists have important things to say about human nature and human potential, we scarcely have a monopoly of wisdom on that topic.

   *Today*: In writing about human potential, I would pay far more attention to differences across cultures and across historical eras.
2. Even within psychology, I had a fairly narrow conception. Almost all of my work on intelligence/potential focused on human cognition—though I had a much broader view of cognition than many of my peers. When I spoke about interpersonal and intrapersonal intelligences, I was fixed on "knowing" about self and others, and not on more affect-laden or personality aspects of human nature. (In this way, my work differs from the well-known work of Daniel Goleman on emotional intelligence.)

   *Today*: I would avoid the almost exclusive valorization of cognition. However, defined or delineated, I would pay more attention to social, emotional, and personality factors. These are often called "noncognitive." I don't endorse that preemptive rejection, but I am content to refer to them as "soft skills."

3. Also, and importantly, my view of human intelligence was pointedly amoral. Any intelligence can be used for benign purposes or malignant purposes: One can use language to write exquisite poetry or to instigate ethnic cleansing.

   *Today:* I have certainly atoned for this omission. For nearly 30 years, with many colleagues, I've been studying the nature of ethical and moral thinking, across the life span, and across many different professions. In this work, we have focused on what it means to use intellectual (and other) strengths in ways that are positive for the wider community. While we recognize that what is "good" can be complex and controversial, we avoid the postmodern trap of refusing to pass judgment on issues of character and behavior.

4. More so than in almost any other line of research on cognition at the time, I searched for relevant evidence from biology—particularly from neuroscience (the representation of different capacities in the human cortex) and from genetics (the extent to which different capacities may be heritable and, if so, to what extent). But in the intervening decades, we have learned a great deal more about the biology of human potential, and so an informed study of the intelligences today would delineate what we do know, what we expect to know, and what remains wrapped in mystery. I suspect that the division into eight or nine intelligences is not sufficiently fine-grained from a neuroscientific point of view, but it captures the insight clearly for educators and nonspecialists.

5. Understandably, in a wide-ranging study of intellect, I focused on the human capacities that pervade the species. Individuals with special talents or specific deficits occupied only a small part of the radar screen. But in the succeeding decades, with colleagues like David Feldman, Lynn Goldsmith, and Ellen Winner, we've taken a much closer look at talent, expertise, prodigiousness, creativity, leadership, and genius, as well as individuals and groups that exhibit flagrant deficits in various areas. Humans differ from one another at least as much as do snowflakes or bacteria—and we need to survey that entire canvas in any account of human potential.

6. Even though the Project on Human Potential was carried out in a school of education, none of the investigators had been trained in that area—we were drawn from the social sciences and the humanities. Accordingly, much of our writing about education was incidental, rather than focal. Both because of my longevity in a school of education and because of the unparalleled interest in my work among educators, I've devoted a great deal of time to thinking about how best to realize human potential—which I would now immediately pluralize as "human potentials."

Any bias that I may have had toward explanations of human potential in terms of genetic contributions has been greatly countered by my (and others') increased knowledge of the large differences in educational outcomes both within and across nations and cultures. Relatively few of those differences seem due strictly to differences in individual human genomes.

Rather, as I have come to put it, how much one achieves within or across intelligences is a joint product of how *important* that capacity is in the society where one happens to live, how highly *motivated* one is to develop that capacity, and how *skilled* are the teachers and the technologies of education available in one's culture—or, nowadays, across the globe.

I have always had lots of curiosity and the desire to explore and understand many facets of life—this is both an advantage (I never get bored) and a challenge (why doesn't Howard stick to one topic?). And so, building on the work on intelligence, I turned my attention to other human potentials, particularly creativity (see Essays 23–24) and leadership (see Essays 25–26). In both cases, borrowing from my undergraduate adviser, the psychoanalyst Erik Erikson, I carried out case studies of exemplary creators and exemplary leaders. These were among the most enjoyable projects that I've undertaken in over a half a century of research and writing.

While I am primarily a social scientist, I have also done extensive work in education. First applying the lens of developmental psychology, I have sought to understand the way that the developing mind comports or clashes with the agenda of schooling—particularly school as it is implemented in modern Western settings. A multiple intelligences perspective can be helpful if there is curricular and pedagogical flexibility; but it's not helpful when there is a laser-like focus on standardized testing, which necessarily narrows what is taught, how it is taught, and how it is assessed. And, focusing on issues that have always been important to me personally, I have contemplated how best to inculcate an appreciation of the true, the beautiful, and the good (see Essays 26 and 28 in *The Essential Howard Gardner on Education*). The subtitle of my book on the topic reveals my slant: "Educating for the Virtues in the Age of Truthiness and Twitter."

Which brings us up to the 21st century.

In what has been termed the era of *homo sapiens,* we have assumed that the species is basically fixed. Once Neanderthals had become extinct, for whatever reason, our ancestors came increasingly to dominate the natural world and to make the planet ours—for better or for worse.

And of course, we have witnessed the better: I would valorize the emergence of religious and moral codes; the efflorescence of the arts; the invention of writing; the domestication of animals; the mastery of farming and hunting; wide access to written materials; the emergence of machines, electronics, and digital technologies—the list goes on.

But, of course, we have also encountered the worse; almost everything just listed has been put not only to good but to malignant use. As an example, religion has motivated murderous crusades while also fostering humane treatment of the old, the lame, and the sick. And, of course, we have had slavery, warfare, genocide, mass pestilence, and lesser forms of chicanery and less flagrant sins. Even globalization—initially lauded as the culmination of the Enlightenment, if not the End of History—can foster ugly forms of tribalism, nationalism, and warfare. And, of course, the use of fossil fuels—long lauded as an expander of human possibilities—risks damaging the planet, if not eliminating life as we know it.

But in recent decades, it has become increasingly clear that *homo sapiens* represents but a chapter in the history of the planet . . . and not necessarily its glorious culmination. From the angle of science and medicine, we can for the first time make significant alterations in the human genome—enhancing or extinguishing traits, or even creating new ones, temporarily or permanently, and enhancing our life span, some say infinitely. Human potential becomes the terrain of what biologists and geneticists can conjure up—and what the rest of the population will allow or even encourage.

From another angle—that of technology, computer science, robotics, and artificial intelligence—we can begin by enhancing (or, again, eliminating) certain already existing human traits. Or, to go further, we can create entities that surpass the species in cognitive capacities (as well as other traits). So much so, indeed, that we may eventually choose or be forced to cede human problem-solving and product-creating capacities to entities that we can only dream about—or that we used to relegate or elevate to the realm of science fiction.

Accordingly, when we change the definition of what it means to be human, or when we create entities—biological or computational—that far exceed what used to be meant by the species and its potential(s), we are likely to need an entirely different set of concepts and explanations.

Indeed, in using the term "we," I am already anticipating that the new species is a linear descendant of ours—but that assumption in itself might be fallacious. Entire lines of evolution have disappeared (where are the dinosaurs or the species that existed before the Cambrian explosion?), and each biological species has its point of origin.

I salute the new experts—the biologists, the geneticists, the chemists, and the computer scientists—who may create new species and new conceptions of posthuman potential. But in doing so, I hope that we do not neglect the stunning achievements of our own species. We did not simply invent writing: We enabled Plato to document what Socrates thought and said; we enabled Shakespeare to portray the enduring features of human beings and Virginia Woolf to capture the experiences of the moment. We did not simply create the first musical instruments and figure out how to write musical scores: We enabled Bach to compose exquisite music, Yo-Yo Ma to perform it on the cello, and Renée Fleming to sing it. And it was human beings who created enduring institutions (churches, schools, and civic offices) and enduring processes (legal systems, constitutions, and Bills of Rights). While a scholarly work necessarily focuses on those findings that can be generalized, human potential as we know it is realized in some way in each and every person. And while we still can, we should admire the heights of human potential . . . and make sure that whatever happens next can preserve and build on those achievements.

## REFERENCES

Gardner, H. (1983). *Frames of mind: The theory of multiple intelligences.* Basic Books.

Gardner, H. (2020). *A synthesizing mind: An intellectual memoir.* MIT Press.

LeVine, R., & White, M. (1986). *Human conditions: The cultural basis of educational developments.* Routledge.

Scheffler, I. (1986). *Of human potential: An essay in the philosophy of education.* Routledge.

# Had I but World Enough and Time

With apologies to English poet Andrew Marvell, I conclude this collection with thoughts on lines of work that I'd like to pursue in the future—and to express the hope that others may also find them worthy of exposition, explanation, and exploration.

As readers of this collection will have noted, I have long been fascinated by issues concerning the mind—and especially the minds of human beings in our time. The essays reveal various facets of my explorations: development, breakdown, intelligence(s), cognition, computation, creativity, leadership, the arts, the humanities, the sciences, morality, and ethics.

## THE SYNTHESIZING MIND

Recently, I have reflected on the human capacity for *synthesizing*. I've sought to understand what's entailed in setting forth a complex issue or set of issues, collecting relevant—and eliminating irrelevant—data, assembling parts and pieces in various ways, publicizing the results, and securing helpful comments and critiques.

It's unrealistic to think that, into my ninth decade, I will be able to solve "the riddle of synthesis" (I first typed "the riddle of *Sisyphus*!") any more than I've dissolved the puzzles of intelligence or of creativity. But I hope to have put the concept of synthesizing sufficiently on the radar screens of reflective scholars, that they will pick up the gauntlet. I am pleased that newly emerging institutions—like the London Interdisciplinary School—organize their curricula around important and perplexing issues like climate change or social inequities or ethics in the era of AI, which call for, indeed *demand*, synthesizing and cross-disciplinary powers.

Of course, as ever more powerful computational devices are created, it will be fascinating and instructive to note and analyze *which* aspects of synthesizing are easily achieved—and in what ways—and *which* continue to require human input, output, reflection, judgment, and final decision.

## GOOD WORK: THE ETHICAL PERSPECTIVE

After decades of focusing on the development and breakdown of human cognition—broadly speaking, in the Piagetian tradition—my colleagues and I now direct our attention to the development of ethical and moral thinking.

We believe that it's important to begin with clear-cut examples of good work. Mihaly Csikszentmihalyi, William Damon, other gifted colleagues, and I first studied over 1,200 admired adult American professionals across the professional landscape. We determined that these individuals exemplified the *three E's*— they were technically *excellent*; they were personally *engaged*; and they sought to carry out work in an *ethical* manner.

While remaining colleagues and friends, members of the original research team went their separate ways. Our own research team at Harvard Project Zero has focused on the development of the habits and the mind of Good Work— what's involved in *becoming the kind of person* who ponders ethical and moral issues; who tries to do the right thing; who seeks to learn from what worked well; and what could have been done differently, and better.

Following our work with adult professionals, we shifted our attention to American colleges and universities. And there we made surprising—and disconcerting—discoveries. Most American students cannot define ethics nor do they think about ethical issues. Even when there is an "ethics center" on campus, they rarely mention it—in sharp contrast to athletic facilities, which almost always garner widespread attention and applause. American students speak of "I" 11 times more frequently than they mention "we." Their parents, alums, and even campus trustees show an even greater focus on "I" than on "we." Our research group has been alarmed by this finding. We have accordingly launched two efforts:

### A. From "I" to "We"

Our team has developed a multifaceted curriculum. Though prepared for secondary school, it has (so to speak) displayed both arms and legs. The curriculum has been used for middle school children as well as at the college level, has been translated into several languages, and has been adopted—and adapted—in countries around the world.

But in our view, training in ethical and moral reasoning and judgment needs to begin much earlier. As just noted, our study of college students in the United States has documented the extent to which the United States in the 21st century is an "I" society, rather than a "we" society. More so than citizens of most other countries, American young people (as well as their parents and friends) are concerned particularly about themselves—their prospects, their futures, their achievements, their place in the social hierarchy.

This egocentrism need not be the case—and studies of other cultures suggest that indeed, it's not necessarily that way elsewhere. And so, in the new line of work entitled "Good Starts," we have been studying the development of a sense of "I" (ego), "we" (the collectivity), and "they" (others, either deemed as neutral or ones from which to keep one's distance or even to disdain) in young children. We strive to understand the decidedly different perspectives of societies like Japan, China, Northern Italy, Scandinavia, and the United States with reference to an emerging sense of self and others . . . and perhaps to achieve a better balance—less egocentrism, more altruism—in our own country.

## B. Educational Innovations Around the World

For a century, American colleges and universities have been widely admired around the world. Indeed, societies ranging from China to Chile have looked to Amherst, Berkeley, Columbia, and Dartmouth as models for their own higher educational systems. And indeed, just as many universities learned from Germany in the 19th century, there has been much to learn from American schools in the past century.

But to focus too much on American examples is misguided—and indeed, as I write in the opening months of 2024, pathologies of American higher education have become all too clear. Accordingly, with my colleagues Wendy Fischman, a longtime researcher at Project Zero, and William Kirby, a historian of modern China, I've launched an effort to learn from promising and successful innovations in other countries. Our first effort is a recently published issue of *Daedalus* (May 2024 https://www.amacad.org/daedalus/advances-and-challenges-interna tional-higher-education)—we hope it won't be the last. Indeed, we have opened a website where those concerned about the quality—and indeed the survival—of higher education in a computer-powerful and consensus-eluding era can converse, exchange, learn, and improve (theworldofhighereducation.squarespace .com).

## INTELLIGENCES

The work described to this point is relatively new. In contrast, I've focused on human intellectual capacities and potentials for over 40 years. To this day, "multiple intelligences" remains the motivation for most of my email correspondence. I've written several books on intelligences and am involved in various educational networks dedicated to this topic.

But my thinking has continued to evolve. I have considered additional intelligences and added one (the naturalist intelligence). I have insisted that it's not sufficient to measure intelligences and characterize individuals with respect to their intellectual profiles. Rather, what's crucial is that we deploy our intellectual powers for positive ends. The wedding of "MI" and "Good Work" is the most important challenge for our time.

## EDUCATION FROM THE CRADLE ON . . . IN THE ANTHROPOCENE

In my efforts to understand and illuminate the essentials of mind, I continue to think about how best to develop our intellectual potential. For most of recorded history, human beings have been considered the principal—if not the sole—focus of educational efforts. But as computational systems become ever-smarter—whether or not they become wiser—it's crucial to map out what such systems *can do* and what they *should* do; what they *may* be able to do in the future; and whether there are *limits or constraints* that can or should be imposed or mandated in "the age of the smart machine."

In the years ahead, I believe that education across the life span will and should be rethought. The established trek—and track—from the three Es to the scholarly disciplines, from homework to preparation for the job market, from mastering the disciplines to adding to the sum of human knowledge—invites reconsideration. We may begin or renew the effort to see education as lifelong; to involve many more individuals as teachers, coaches, masters, mentors, and fellow learners; to draw on a wide range of society institutions and facilities; and to rethink curricula. We need to reconsider what all of us *should* know and be able to do, what some of us *want* to know and be able to do, and what many of us would *like* to do vocationally and avocationally—and how to achieve these goals.

Your kind attention: If you want to follow these lines of work, there's an easy thing to do for the time being. Just go to my website, www.howardgardner.com, and scan the ever-growing set of blogs—currently several hundred and still going strong.

## GRATITUDE

To this point in the concluding essay, I've focused on the work that I've done—mostly with gifted colleagues, at times in a more solitary way. And since, by the clock, I spend the majority of my hours at work, that seems fine.

But there are two realms that loom very large in my life—my work family and my life family.

For nearly 60 years, I have worked at Harvard Project Zero (HPZ), a research and development entity at the Harvard Graduate School of Education (see Essay 3 and see *The Essential Howard Gardner on Education*). For many years, David Perkins and I co-directed HPZ. It grew from a small set of researchers carrying out psychological research on the arts to a large research group, with over a dozen "principal investigators," who at any one time are carrying out between 20 and 30 research projects on many facets of education around the world (see https://pz.harvard.edu).

When I turned 80, I resigned from the leadership team of which I was the nominal leader, believing it was high time for others to oversee HPZ. I laid out a set of principles that I hope will govern the organization in the period ahead.

But I keep attending HPZ meetings and, when asked, offer my advice and help. My wife, Ellen Winner, whom I met at Project Zero over 50 years ago, is currently carrying out a comprehensive study of the impact of HPZ ideas all over the world. Stay tuned for her findings, which I believe will demonstrate the considerable value and impact of our work.

Which leaves family—that's the most important to me: My wife, Ellen, our children, and our (by now) five grandchildren—all cherished. I think about them all the time, every day—admire and applaud their achievements, try to be helpful when I can be, and am deeply moved by their dedication to our family across the generations. We have been blessed with good health and good luck. When I turned 80, I wrote, signed, and shared an "ethical will" with our grandchildren—I hope that they will be guided by these principles in the years ahead and will, in turn, pass them on to their generations, just as my parents and grandparents passed on their values to me.

In this volume, *On Mind*, and in the accompanying volume, *On Education*, I have shared my thoughts, my findings, my beliefs, and my values, with colleagues of many interests and in many places. It's my hope that one, or 10, or even more of those of you who are reading these words will be stimulated to reflect on, to critique, and to build upon these ideas, findings, and recommendations. We are not related in a family way, but we are part of the most important entity I know: the human family, which I pray will become a humane enterprise as well.

Howard Gardner
Spring 2024

# Original Publication List

I am extremely grateful to the various publications—and to my several co-authors—who have kindly granted permission to reprint articles and chapters that were written over the last half century—or, indeed, longer! Scholarship is a collaborative project—as is publishing—and I have been very fortunate with my colleagues, many of whom have become lifelong friends.

Except for corrections of errors or necessary updates (e.g., of dates!) and elimination of text that appears elsewhere in the volume (e.g., the list of multiple intelligences), the wording of articles and chapters remains mostly unchanged.

It is important that all users of this book be able to access the original publications, as they appeared. Accordingly, the list below includes URLs to all the materials in the book, including any articles, books, URLs that are cited therein. Any source that is discussed substantively is also cited in this publication.

The papers in these/this volume were written more than half a century ago—from 1970 to the present. Needless to say, I would not say exactly the same things in exactly the same way in 2024 as I did decades ago. In this collection, I have followed a few guidelines. I only included papers that I believe are of biographical and/or scholarly significance. When there were clear factual errors, I sought to correct them as much as possible. When there were clear updates, I included them. But otherwise, the texts are faithful to the original expression.

There are important issues concerning examples and wording. Without doubt, were I to rewrite these papers today, I would use more timely examples and I would also avoid using words like *man, he, his* generically—and I would delete concepts and phrases that are no longer used, such as certain medical and diagnostic categories that are now considered belittling or anachronistic. I have updated the text in some ways but still have retained the aura of the original. I know these decisions won't satisfy everyone—they don't even satisfy me!—but I have tried to do my best.

I could not have carried out any of this work without having wonderful colleagues, assistants, and students. I have sought to acknowledge their exemplary contributions and I apologize in advance to anyone who has been inadvertently left out. If there are future editions of these collections, I pledge to make all corrections.

# REFERENCE LIST

## From "Influences"

Gardner, H. (1980). Jean Piaget: The psychologist as Renaissance man. *New York Times*, Section 4. https://www.nytimes.com/1980/09/21/archives/jean-piaget-the-psychologist -as-renaissance-man-the-philosophers.html

Gardner, H. (2017). Jerome Seymour Bruner necrology. Proceedings of the American Philosophical Society. Penn Press.

Gardner, H. (2000). Project Zero: Nelson Goodman's Legacy in Arts Education. *The Journal of Aesthetics and Art Criticism, 58*(3), 245–249. https://doi.org/10.2307/432107

Gardner H. (1997). Creative genius. In S. Schachter, & O. Devinsky (Eds.), *Behavioral neurology and the legacy of Norman Geschwind* (pp. 47–51). Lippincott–Raven.

## From "Early Work"

Gardner, H. (1970). Piaget and Lévi-Strauss: The quest for mind. *Social Research, 37*, 348–365. https://www.jstor.org/stable/40970021

Gardner, H. (1970). From mode to symbol: Thoughts on the genesis of the arts. *British Journal of Aesthetics, 10*, 359–375. https://doi.org/10.1093/bjaesthetics/10.4.359

Gardner, J., & Gardner, H. (1970). A note on selective imitation in a six-week old infant. *Child Development, 41*, 911–916. https://doi.org/10.2307/1127349

Gardner, H. (1970). Children's sensitivity to painting styles. *Child Development, 41*(3), 813–821. https://doi.org/10.2307/1127226

## From "Developmental Psychology"

Gardner, H. (1972). Style sensitivity in children. *Human Development, 15*, 325–338. https://doi.org/10.1159/000271255

Winner, E., Rosenstiel, A. K., & Gardner, H. (1976). The development of metaphoric understanding. *Developmental Psychology, 12*(4), 289–297. https://doi.org/10.1037 /0012-1649.12.4.289

Gardner, H. (1981). Intimations of artistry. In R. Stavy (Ed.), *U-shaped behavioral growth*. Academic Press.

Gardner, H. (1979). Developmental psychology after Piaget: An approach in terms of symbolization. *Human Development, 22*(2), 73–88. https://www.jstor.org/stable/26764784

## From "Introduction to the Study of Brain Damage"

Gardner, H. (1973). The contribution of operativity to naming capacity in aphasic patients. *Neuropsychologia, 11*(2), 213–220. https://doi.org/10.1016/0028-3932(73) 90010-9

Zurif, E., & Gardner, H. (1975). Bee but not be: Oral reading of single words in aphasia and alexia. *Neuropsychologia, 13*(2), 181–190. https://doi.org/10.1016/0028 -3932(75)90027-5

Gardner, H., & Winner, E. (1977). The comprehension of metaphor in brain-damaged patients. *Brain, 100*(4), 717–729. https://doi.org/10.1093/brain/100.4.717

Gardner, H. (1994). The stories of the right hemisphere. *International Schools Journal: Nebraska Symposium on Motivation, 41,* 57–69. PMID: 7537866

## From "Introduction to Multiple Intelligences"

Gardner, H. (2006). Chapter 1: In a nutshell. From *Multiple intelligences: New horizons in theory and practice revised*. Basic Books (pp. 3–24).
Gardner, H. (2022). A "smart" lexicon. *Roeper Review.* 44(2), 82–84. https://doi.org/10.1080/02783193.2022.2043504
Gardner, H. (1983). Artistic intelligences. *Art Education, 36,* 47–49.
Gardner, H. (1999). Who owns intelligence? *The Atlantic.* http://arowe.pbworks.com/w/file/fetch/51366150/Who%20Owns%20Intelligence.pdf

## From "Cognition"

Gardner, H. (1987) Definition and scope of cognitive science. From *The Mind's New Science*. Basic Books (pp. 5–7).
Gardner, H. (1992). Scientific psychology: Should we bury it or praise it? *New Ideas in Psychology, 10*(2), 179–190. https://doi.org/10.1016/0732-118X(92)90027-W

## From "Heights of Cognition: Creativity"

Gardner, H. (1993). Seven creators of the modern era. From *Creativity* (Ed. J. Brockman). Simon & Schuster (pp. 28–47).
Weinstein, E., & Gardner, H. (2018). Creativity: The view from big C and the introduction of tiny c. In R. J. Sternberg & J. C. Kaufman (Eds.). *The nature of human creativity* (pp. 94–109). Cambridge University Press.

## From "Leadership"

Gardner, H. (1998). Leadership 1. K-12 education: perspectives on the future (5–24). Based on a presentation to the Van Andel Educators Institute, August, 1997. The Van Andel Education Institute.
Gardner, H. (2020). On good leadership: Reflections on leading minds after 25 years. https://www.multipleintelligencesoasis.org/blog/2020/9/21/on-good-leadership-reflections-onnbspleading-minds-after-25-years
Gardner, H. (2006). Mental representations: The 80/20 principle. In *Changing minds* (pp. 7–18). Harvard Business Review Press.
Gardner, H. (2008). *What is good work?* Lecture delivered as part of The Tanner Lectures on Human Values at the University of Utah.
Gardner, H. (2008). *Achieving good work in turbulent times*. Lecture delivered as part of The Tanner Lectures on Human Values at the University of Utah.

## From "The Professions"

Gardner, H. (2005). Compromised work. *Daedalus*, *134*(3), 42–51. https://www.jstor.org
    /stable/20027997

Horn, L., & Gardner, H. (2006). The lonely profession. In S. Verducci & W. Damon
    (Eds.), *Taking philanthropy seriously: Beyond noble intentions to responsible giving*
    (pp. 77–93). Indiana University Press.

Gardner, H. (2015). Reclaiming disinterestedness for the digital era. In D. Allen & J. S.
    Light (Eds.), *From voice to influence: Understanding citizenship in a digital age.*
    University of Chicago Press.

Gardner, H. (2016). Is there a future for the professions?: An interim verdict. *The Hedge-
    hog Review*, *18*(1). https://hedgehogreview.com/issues/work-in-the-precarious-eco
    nomy/articles/is-there-a-future-for-the-professions-an-interim-verdict

## From "Minds of the Future"

Gardner, H. (2009). The five minds for the future: Cultivating and integrating new ways of
    thinking to empower the education enterprise. *The School Administrator Magazine*,
    *66*(2), 16–21.

Gardner, H. (2020). Understanding the synthesizing mind. From *A Synthesizing Mind*
    (pp. 211–236). MIT Press.

Gardner, H., & Davis, K. (2013). Why the app generation? In *The app generation: How
    today's youth navigate identity, intimacy, and imagination in a digital world.* Yale
    University Press.

## From "Miscellanea"

Gardner, H. (2020). Of human potential: A 40-year saga. *The Association for the Gifted:
    Council for Exception Children 43*(1). https://doi.org/10.1177/0162353219894406

# Index

# Permissions

# About the Author

*Howard Gardner* is the Hobbs Research Professor of Cognition and Education at the Harvard Graduate School of Education. He is a leading thinker on education, human development, and cognition. In 30 books, translated into over 30 languages, Gardner has written extensively about intelligence, creativity, leadership, and professional ethics. He has received honorary degrees from 31 colleges and universities, including institutions in Bulgaria, Canada, Chile, Greece, Hong Kong, Ireland, Israel, Italy, South Korea, and Spain. A MacArthur Prize Fellow in 1981, he is the winner of the 1990 Grawemeyer Award in Education, the 2011 Prince of Asturias Prize in Social Science, the 2015 Brock International Prize in Education, and the 2021 AERA Distinguished Contributions to Research in Education Award.

In recent years Gardner has completed a national study of higher education; *The Real World of College,* co-authored by Wendy Fischman, appeared in 2022. Gardner also directed an international study of the United World Colleges, a network of secondary schools, and, with Wendy Fischman and William Kirby, has co-edited a collection on innovations in higher education around the world (https://www.amacad.org/daedalus/advances-and-challenges-international-higher-education). Other recent books include *Good Work; Changing Minds; The Development and Education of the Mind; Multiple Intelligences: New Horizons; Truth, Beauty, and Goodness Reframed;* and *The App Generation,* co-authored with Katie Davis. His intellectual memoir, *A Synthesizing Mind,* was published in 2020. A *Festschrift* marking Gardner's 70th birthday is available online at www.howardgardner.com. A regular, committed blogger, Gardner's current thinking about good work, multiple intelligences, and synthesizing can be followed on his website.